ScienceWorld 7

Australian Curriculum edition

Lesley Englert · Peter Stannard · Ken Williamson
BA, DipT · BSc, DipEd · BSc (Hons), DipEd

This edition published in 2021 by

Matilda Education Australia, an imprint
of Meanwhile Education Pty Ltd
Melbourne, Australia
T: 1300 277 235
E: customersupport@matildaed.com.au
www.matildaeducation.com.au

First edition published in 2012 by Macmillan Science and Education Australia Pty Ltd

Publication data

Author: Lesley Englert, Peter Stannard and Ken Williamson
Title: ScienceWorld 7 Workbook: Australian Curriculum edition
ISBN: 978 1 4202 2994 3

Publisher: Peter Saffin
Project editor: Inge Note
Illustrator: Chris Dent; Andy Craig and Nives Porcellato
Cover designer: Dim Frangoulis
Text designer: Firefly Education
Production control: Loran McDougall
Typeset in Helvetica, Trade Gothic and Linotype Kaliber by Firefly Education
Cover images: Lonely Planet/Martin Cohen; Shutterstock/Silver-John

Printed in China by Central
Sep-2024

About this book

This workbook is full of exercises which will help you understand the material in the textbook, think about the information, relate it to your own experiences, and write about it in your own words.

The exercises ask questions and pose problems to give you the opportunity to reflect on your knowledge, then show what you know.

Do you find that you know the answer but often do not know how to write it in the correct form? The workbook teaches you how to organise this information into many different written and oral forms.

On the following page you will see some sample pages from the workbook. You will also see descriptions of the various features in the workbook. These descriptions will help you become familiar with the workbook.

Getting to know your workbook

Each chapter begins with the same chapter number and name as found in your textbook.

Sometimes you will see a Challenge box. These boxes contain challenging questions or problems that you might like to attempt.

Chapter 6

Separating mixtures

Overview

Suspensions
solid + liquid where the solid settles out if left to stand

Separated by
- decanting
- using a centrifuge
- filtering

Coloured substances

Separated by
- chromatography

Mixtures

Solutions = solvent + solute
concentrated, dilute, solubility

Separated by
- evaporation
- distillation

Solids

Separated by
- dissolving one
- magnet
- gravity
- froth flotation

What do you know already?

Exercise 1 Problem-solving

Problem: Imagine you are a farmer and there is a severe drought. Your sheep desperately need water, and the only water is bore water which is too salty and muddy.

How can you produce fresh water from the bore water?

Suggest a possible solution on the next page.

Chapter 6 – Separating mixtures 978 1 4202 2994 3 43

This is called the Overview—it is a diagram of the way the ideas in the chapter are connected.

The section of the workbook called During reading contains exercises which ask you questions about the ideas on certain pages of your textbook.

The first section of the chapter is called What do you know already? This will show how much you already know about the work in the chapter. This is not a test.

During reading

Exercise 2 Sketching and labelling

After reading page 57

An inference is an explanation you can give based on your observations. The inference may or may not be true.

Alan is riding his bicycle up a steep hill at high speed.

Your task is to sketch and label four arrows to show:
- muscle force
- gravitational force
- frictional force
- air resistance force.

The arrows you sketch should show:
- the direction of the force
- the relative strength of the force.

Exercise 3 Using data tables and graphs

After reading Measuring forces on page 58 and doing Investigation 8 on pages 61–62

Matty decides to check whether friction is greater for heavy objects than for light objects. He uses a spring balance to pull a wooden block across the desk. Then he pulls two blocks, one on top of the other, then three, then four. Here are his results.

Number of blocks	Force needed to pull blocks (newtons)			
	Trial 1	Trial 2	Trial 3	Average
1	3.1	3.3	3.2	
2	6.7	6.4	6.4	
3	9.8	9.5	10.0	
4	12.8	12.6	13.1	

To calculate the average, add the results for the three trials and divide by 3.

1 Complete Matty's data table by calculating the averages.

Chapter 3 – Forces 978 1 4202 2994 3 19

The Exercises contain questions for you to answer. Sometimes they require you to write words in the space provided. Others ask you to draw data tables or graphs. Some ask you to create stories or to solve problems.

Chapter Roundup will check how much you know about the main ideas in the chapter. Be prepared to write answers, teach someone else, have a group discussion, or role-play.

Chapter 1
Working in a laboratory

Overview

An overview is a diagram of the way the main ideas in the chapter are connected. Each of the chapters in this workbook begins with an overview.

The four large arrows in the diagram below are the four main headings in the chapter.

Laboratory equipment

Safety in the laboratory

Using a burner

Science is investigating

Investigations

Exercise 1 Listing

List the ten safety rules that are being broken in the illustration above.

1 ______________________

2 ______________________

3 ______________________

4 ______________________

5 ______________________

6 ______________________

7 ______________________

8 ______________________

9 ______________________

10 ______________________

Exercise 2 Making and justifying decisions

Read each of the statements below. Circle TRUE if you think the statement is true, or FALSE if you think it is false. Be prepared to give reasons for your answer.

1 A science laboratory is like a kitchen, only more hazardous. TRUE FALSE

2 Normal-sized test tubes hold more water than beakers hold. TRUE FALSE

3 A measuring cylinder is used for heating liquids. TRUE FALSE

4 The Bunsen burner was named after a German scientist called Robert Bunsen. TRUE FALSE

5 The yellow flame of a Bunsen burner is hotter than the blue flame. TRUE FALSE

6 You should tie back long hair whenever you are working in a laboratory. TRUE FALSE

7 Goggles are worn in the science laboratory to reduce the glare from burner flames. TRUE FALSE

8 Minor burns can be treated with cold running water. TRUE FALSE

9 When you boil sea water, the water evaporates leaving a small amount of salt. TRUE FALSE

10 Science is doing experiments. TRUE FALSE

Keep your answers in mind as you read the textbook.

During reading

Exercise 3 Labelling

After reading Laboratory equipment on pages 3–4

Trying not to refer to your textbook, use the following words to label the laboratory equipment in the box below.

beaker	funnel	metal tongs	test tube holder
Bunsen burner	gauze mat	spatula	test tube rack
conical flask	glass stirring rod	stand and clamp	tripod
dropper	measuring cylinder	test tube	watch glass

Exercise 4 Recalling information

After reading Safety in the laboratory on pages 8–10

1 What do the following labels mean?

a 

____

b 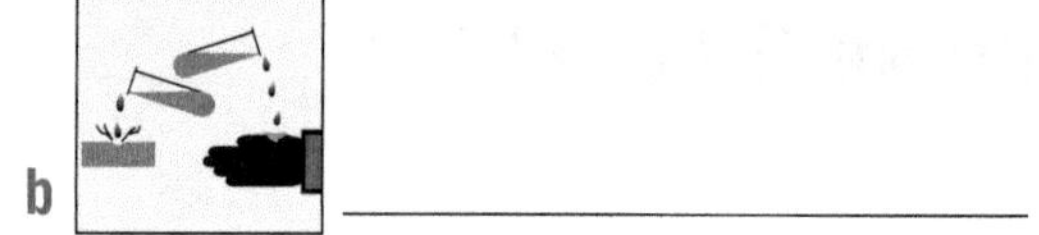
____

2 Sketch the labels for chemicals which are:

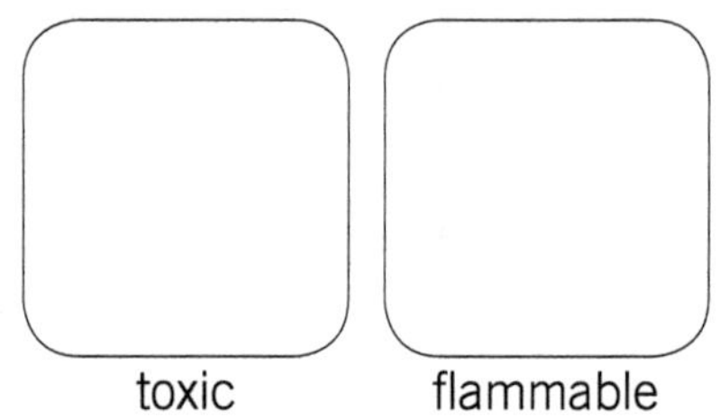

3 Select the words from the list on the right to complete the sentences below.

a Never ________________ any chemicals in the laboratory.

b ________________ all breakages to your teacher.

c If you spill ________________ on your skin, wash immediately with plenty of ________________, then tell your ________________.

d Take care not to ________________ anyone who is using chemicals.

e Do not ________________ in the laboratory.

f Never point a ________________ towards your own face or anyone else's.

g A ________________ can be wrapped quickly around someone to smother flames.

h Know what you are doing in the laboratory. Read the ________________ carefully.

i You should never put oil, petrol or ________________ liquids down the sink.

j Use a ________________ mat so the benches are not scorched during heating.

bump
chemicals
corrosive
eat food
fire blanket
heatproof
instructions
report
taste
teacher
test tube
water

Exercise 5 Labelling

After reading Using a burner on page 11 and doing Investigation 1

1 Neatly print the five labels below in the correct boxes in the diagrams on the right.

air hole open
air hole closed
yellow flame
blue flame
hottest part of flame

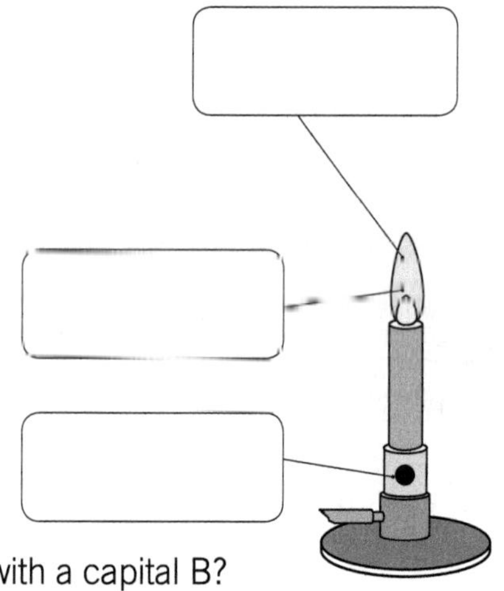

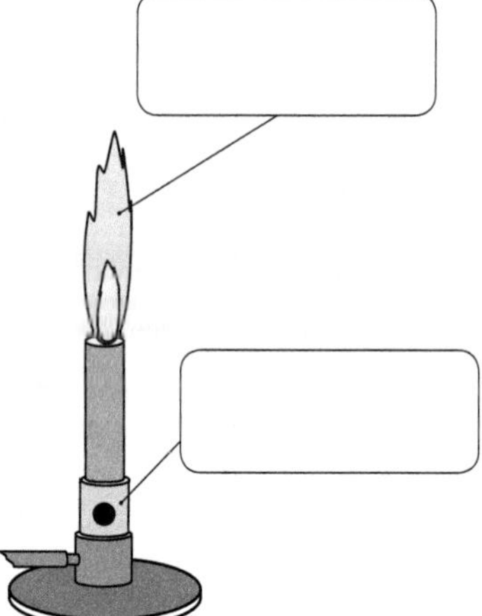

2 Why does 'Bunsen' in Bunsen burner start with a capital B?

____

Exercise 6 Choosing the correct sequence

After doing Investigation 1 Part C on page 13

1 The diagram on the right shows the apparatus needed to heat a liquid in a beaker. Here is a list of the four steps (not in the correct order).

a Light the burner.

b Open the air hole on the burner and turn down the gas.

c Place the beaker on a gauze mat.

d Rotate the collar on the burner so that it is closed.

Tick the alternative below which shows the correct sequence of steps for this activity.

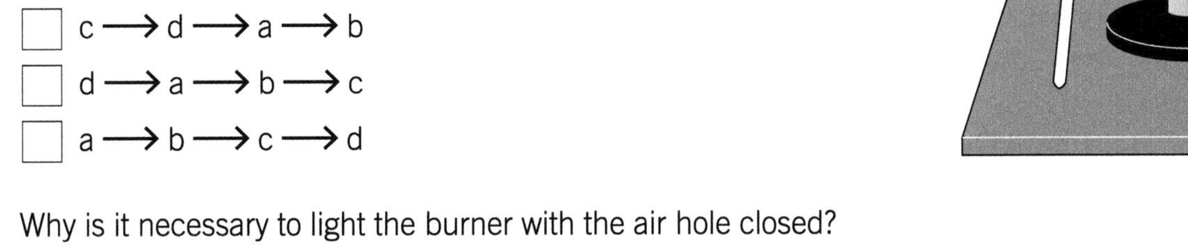

☐ b ⟶ a ⟶ d ⟶ c

☐ c ⟶ d ⟶ a ⟶ b

☐ d ⟶ a ⟶ b ⟶ c

☐ a ⟶ b ⟶ c ⟶ d

2 Why is it necessary to light the burner with the air hole closed?

____

____

Exercise 7 Using sequence linking words

When actions occur one after the other, *sequence linking words* are used to describe the order in which they occur.

Sequence linking words (for actions that occur one after the other)

first, second	then	before	finally	subsequently
to begin with	later	next	gradually	initially
when	from there	last	initial	after

1 Underline the *sequence linking words* in the following paragraph.

> Before lighting a Bunsen burner, place it on a heatproof mat, then connect the gas hose to the gas tap. Next, rotate the collar so that the air hole is closed. Finally, light a match, then turn on the gas and gradually bring the match close to the top of the barrel.

2 Use *sequence linking words* to complete the correct sequence of actions to respond to a fire in the laboratory.

Initially, you should stay calm and ____________________________

____

____

____

Exercise 8 Writing instructions

1 Look at the instructions a, b, c, d in Exercise 6 on the previous page.
List the first word in each instruction.

________________ ________________ ________________ ________________

Look at the first three instructions for Lighting the burner in Part A on page 12 of your textbook.
List the first word in each instruction.

________________ ________________ ________________

What do all these words have in common?

Did you realise that they are all *action words*—words which tell you to do something?
These action words are known as **verbs**.

2 Circle all the verbs in the list below. If the whole sentence makes sense, then the word is a verb.
If you are not sure, try to place the word in this sentence:

I can ________________ something.

jump	run	label	wash	learn	activities	separate
fold	filter	boiling	dangerous	learner	very	too
explosion	firmly	open	often	remove	crush	gently

3 As well as starting with a verb, instructions are usually written as a list of steps. Each step is written on a separate line and the order is shown by using numbers: 1, 2, 3, 4 ... Complete the instructions below.
Make sure each instruction starts with a verb, and *underline* the verb.

Instructions for washing a dog	Instructions for what to do if there is a fire (page 9 of your textbook)
1 Collect shampoo, towels and a tub.	
2 Fill the tub with water.	
3	
4	
5	
6	

Exercise 9 Writing a report

After reading Writing reports on pages 14–16

Study the report of an experiment and answer the questions below.

Experiment 4. 21 November

STOPPING DISTANCE

AIM: To find out what effect the mass of a truck has on its stopping distance.

METHOD: We set up a ramp as shown and used a metre rule to measure how far the empty truck travelled before stopping.

Ramp

stopping Distance

We then added metal washers to the back of the truck to increase its mass, and measured the stopping distance again.

RESULTS:

Mass added (No. of washers)	stopping Distance (metres)
0	0.6
1	0.9
2	1.1
3	1.3
4	1.4

CONCLUSION: The heavier the truck is, the longer it takes to stop.

1 Re-write the aim of the experiment as a question. ______________________

____

____

2 List the materials used in the experiment. ______________________

____

3 How did the students change the mass of the truck? ______________________

____

4 The conclusion of the experiment is usually a **generalisation**. It could also be written as:

Heavy trucks ______________________ than light trucks.

5 The report did not include a discussion. What do you write in a discussion? ______________________

____

____

Exercise 10 Being specific

After reading Skills for investigating on page 16

1 When making observations it is important to be accurate and specific, giving details and measurements where possible. Make the observations below more specific. The first one is done for you.

a It is cold today. The temperature today is 15 degrees Celsius.

b My friend lives in a house down the road a bit. ______

c We caught a few whiting. ______

d We won most of our games last year. ______

______

______

______

2 After doing Investigation 3 on pages 17–18, make the observations below more specific, describing in detail what happened.

You may need to refer to your notes, and discuss your observations with other students. Make sure you use complete sentences, starting with a capital letter and finishing with a full stop.

PART A

Went cloudy. ______

______

______

PART B

It bubbled. ______

______

______

PART C

The test tube felt funny ______

as there was a reaction. ______

______

Exercise 11 Writing sentences

When you speak, you don't always use complete sentences. However, when you are doing exercises it is important to use complete sentences. A sentence is a fully formed thought which must contain a verb and, of course, should always begin with a capital letter and end with a full stop, question mark or exclamation mark.

Use the words below to form complete sentences.

1 going shopping ______________________________

____

2 hair tied back ______________________________

____

3 because he was tired ______________________________

____

4 diagram too small ______________________________

____

5 it worked water collected in the beaker ______________________________

____

Roundup

Exercise 12 Doing the exercises in the textbook

At the end of each section in your textbook, there is a set of questions called **Check**. To answer these, you need information from the section just before the questions. Also, some questions are more difficult than others. Generally, the questions at the beginning of the set are easier than those towards the end.

1 Find Check 1 on page 19. This is a simple question which you should be able to do quickly using information straight from the textbook.

Complete the caption telling Brooke how to do Check 1.

2 Look at Check 9 on page 20. This is a more difficult question where you have to work it out for yourself. Discussing it with others often helps.

Complete the captions at the top of the next page by adding the missing words.

The first thing I did in the experiment was to put the tripod on a heat-proof mat. Then secondly, I ______

Obviously c – filling the beaker with water – comes before ____ and d comes before ____

And the last thing was to put the burner under the gauze mat. So, the order is: f ⟶ a ⟶ ______

3 For more difficult questions you have to read them very carefully, then apply the information in the book. Help Quinn do Challenge 1 on page 21 by completing the captions.

Exercise 13 Using the glossary

1 In your own words write down the meaning of these two words:

laboratory ______________________________

apparatus ______________________________

Now check the Glossary (pages 276–279) and compare the authors' meaning with yours.

2 In the Glossary, laboratory has its pronunciation given as la-BOR-a-tory. Why is the BOR syllable in capitals?

____

Make sure you can pronounce la-BOR-a-tory correctly.

Chapter 2

Science skills

Overview

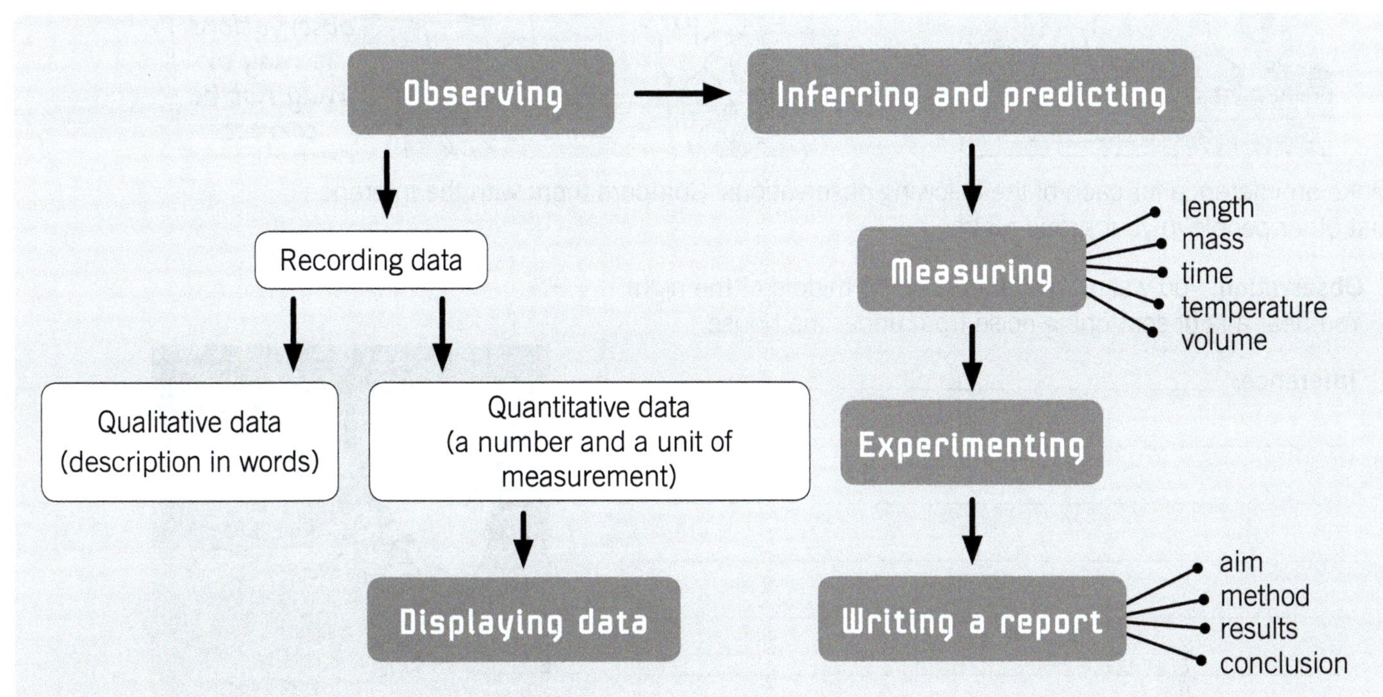

What do you know already?

Exercise 1 Completing a table

Complete the table below which describes how you measure quantities such as length.
The first one is done for you.

You measure ...	Using a ...	The measurement is in ...
length	ruler	centimetres
time		seconds
volume	measuring cylinder	millilitres
	balance	
		degrees Celsius

Exercise 2 Observing and inferring

After reading Inferring and predicting on pages 27–28

An inference is an explanation you can give based on your observations. The inference may or may not be true.

Make an inference for each of the following observations. Compare them with the inferences that other people in your group make.

1 **Observation**: You wake up suddenly in the middle of the night. You hear a faint scratching noise from under the house.

Inference: ______________________________

____

____

____

____

____

2 **Observation**: The leaves on the tree have many holes in them.

Inference: ______________________________

____

3 **Observation**: There is broken glass at the intersection.

Inference: ______________________________

____

4 **Observation**: When we added acid to the zinc it bubbled vigorously.

Inference: ______________________________

____

5 **Observation**: The drop of methylated spirits on my hand disappeared.

Inference: ______________________________

____

6 **Observation**: This rock outcrop on the surface of Mars looks like a human face when lit from the side (photos on page 27).

Inference: ______________________________

____

Exercise 3 Predicting

After reading Inferring and predicting on pages 27–28

Complete the table below.

Observations	Predictions
1 Mickey bought a hot dog for lunch on Monday. 2 Mickey bought a hot dog for lunch on Tuesday. 3 Mickey bought a hot dog for lunch on Wednesday.	
1 Ants are trekking into the house. 2 The humidity is nearly 100%. 3 The sky is grey and overcast.	
1 2 3	There will be a full moon tomorrow night.

Exercise 4 Completing a table

After reading Measuring on pages 32–33

Place the following observations in the correct column in the table.

- The flower is drooping.
- She weighs 42 kg.
- Ian Thorpe has big feet.
- That carton holds 1 litre.
- My hair is shorter than Becky's hair.
- He is 172 cm tall.
- The temperature is 28°C.
- I feel very healthy.

Qualitative observations	Quantitative observations

Exercise 5 Reading a scale

After reading Measuring on pages 32–33

1 Read the ten scales below. Score your answers out of 10. A score of more than 7 is good. For any you get wrong, go back and make sure you know why you are wrong.

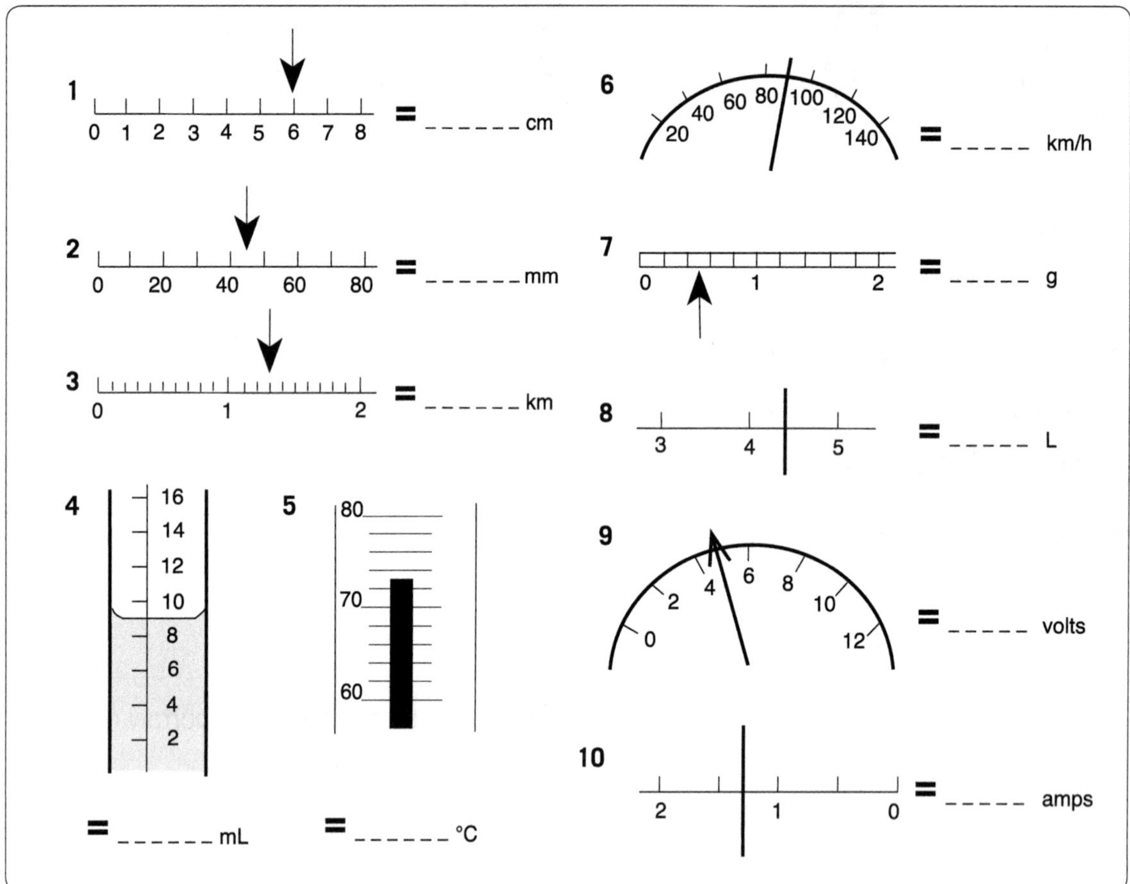

2 Every measurement must have a unit after it, for example 5 cm, where *cm* is the symbol for centimetres. Write the full name for the following symbols. Make sure your spelling is correct.

a mm ______________________

b s ______________________

c mL ______________________

d g ______________________

e °C ______________________

f L ______________________

g km/h ______________________

h amps ______________________ (You may need to use a dictionary for this one.)

Exercise 6 Interpreting data tables

Measurements or observations are called **data**. The best way to record data in a form that is easy to understand is to use a **data table**.

After he reached the age of two, Paul's parents measured his height every year. They recorded these measurements in a data table.

Age	Height (cm)
2	88
3	92
4	94
5	94
6	97
7	101

1 How tall was Paul when he was four?

____

2 How old was he when he was 92 cm tall?

____

3 How old was he when he reached 1 m in height?

____

4 Predict how tall he will be when he is eight.

____

Exercise 7 Graphing

After reading Displaying data on page 40

A graph is a way of displaying data.

Look at the data table for Paul's height from the ages of 2 to 7.

1 Display this data as a line graph. The first point has been done for you.

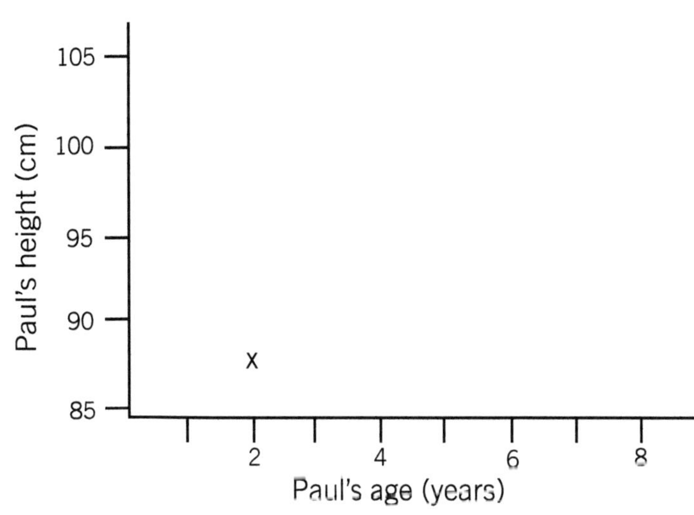

2 What pattern can you see in the line graph?

____

____

____

3 Which measurement doesn't seem to fit this pattern?

____

4 Write an inference to explain the measurement which doesn't seem to fit.

____

____

____

Exercise 8 Generalising

Yup, yup, I'm one smart puppy. Tee hee, woof.

In your dreams, pea brain.

1 Underline the word(s) in each statement which indicates that the statement may be true but not necessarily in all cases.

- Most crustaceans live in water.
- Cats are usually more intelligent than dogs.
- White clothes are normally cooler than darker clothes.
- Black clouds generally indicate rain.
- As a general rule, women are better drivers than men.

Generalisations can be written in three different ways:

- as a statement, e.g. **Heavy people go faster down a water slide**.
- as a comparison, e.g. **The heavier people are, the faster they will travel down a water slide**.
- using *generalisation linking words*, e.g. **As a general rule, heavy people travel faster down a water slide than lighter people**.

Generalisation linking words (for actions that occur one after the other)

generally	usually	commonly	many
normally	as a general rule	most	mostly

2 Write four generalisations using some of the *generalisation linking words*.

a ____________________

b ____________________

c ____________________

d ____________________

Roundup

Exercise 9 Observing, inferring, predicting, generalising

1 Write an **observation** for this cartoon.

2 Write an **inference** for this cartoon.

3 If he received a ticking parcel every day for a week, what **prediction** could he make?

4 What **generalisation** could he make?

Exercise 10 Analysing a report

As an assignment, Amy's teacher asked her to design and carry out an experiment. The experiment was to find out the effect of liquid fertiliser on the growth of plants.

On the right is Amy's unfinished report.
Read it carefully, then answer the questions below.

1 Look at Amy's aim. Is this what her teacher asked her to investigate? If you were given this assignment, how would you write your aim?

2 Rewrite Amy's method to improve the design of the experiment.

3 Is Amy's experiment a fair test of the effect of liquid fertiliser on radish plants? Explain your answer.

4 Amy hasn't written her conclusion. Write one for her.

<u>Using Fertiliser</u>

Aim: To see how tall plants can grow.

Hypothesis: When fertiliser is added to plants they grow taller.

Method:
1 I got 2 pots and filled them with potting mix.
2 In each pot I planted 3 radish seeds.
3 I watered the pots every day with 250 mL of water containing 2 drops of liquid fertiliser.
4 Once the seeds germinated I put a numbered tag on each seedling. Then I measured their height every 3 or 4 days.

Results:

Plant No.	Height (cm)						
	3rd July	6th July	10th July	13th July	17th July	20th July	24th July
1	1.0	3.2	5.6	8.9	11.6	13.2	14.3
2	1.1	2.9	5.4	8.9	11.3	12.2	12.4
3	0.9	2.9	5.1	8.4	11.2	11.9	12.0
4	1.2	3.3	5.9	9.2	12.2	13.6	14.8
5	0.8	2.6	3.1	3.1	This plant died		
6	1.0	3.0	5.3	9.1	11.0	11.1	11.1

Conclusion:

Chapter 3
Forces

Overview

What do you know already?

Exercise 1 Problem-solving

Problem 1 You are helping to build a pyramid and have to move a 200 kg stone block a distance of 100 metres. Any ideas?

Write down three possible solutions.

1 ______

2 ______

3 ______

Problem 2 Bozo has agreed to jump off a 20-storey building. How can he make the jump without killing himself? Write down a possible solution.

______

______

______

Keep your solutions to these two problems in mind as you read and work your way through the chapter. This way you will find out which principles of physics you already know.

During reading

Exercise 2 Sketching and labelling

After reading page 57

An inference is an explanation you can give based on your observations. The inference may or may not be true.

Alan is riding his bicycle up a steep hill at high speed.

Your task is to sketch and label four arrows to show:

- muscle force
- gravitational force
- frictional force
- air resistance force.

The arrows you sketch should show:

- the direction of the force
- the relative strength of the force.

Exercise 3 Using data tables and graphs

After reading Measuring forces on page 58 and doing Investigation 8 on pages 61–62

Matty decides to check whether friction is greater for heavy objects than for light objects. He uses a spring balance to pull a wooden block across the desk. Then he pulls two blocks, one on top of the other, then three, then four. Here are his results.

Number of blocks	Force needed to pull blocks (newtons)			
	Trial 1	Trial 2	Trial 3	Average
1	3.1	3.3	3.2	
2	6.7	6.4	6.4	
3	9.8	9.5	10.0	
4	12.8	12.6	13.1	

To calculate the average, add the results for the three trials and divide by 3.

1 Complete Matty's data table by calculating the averages.

2 Draw a line graph of the data Matty gathered. The axes have been drawn for you.

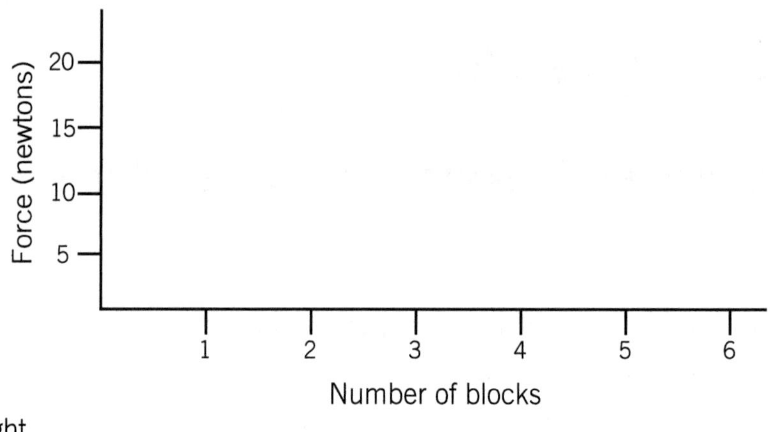

3 Summarise Matty's results by completing the following sentences. The force needed to pull two blocks is about twice the force needed to pull ____________________; and the force needed to pull three blocks is about ____________________ the force needed to pull one block. So, as the weight increases, so does the ____________. Or, the greater the weight, the ______________________________________.

4 Matty would like to know what the friction is for six blocks, but he only has four blocks.

Use your graph to predict what the friction would be for six blocks. ____________ newtons.

Exercise 4 Recalling and justifying

After reading Frictional Forces on pages 61–62

1 Complete this generalisation.

When you try to push a heavy object, the force acting against you is called ____________________.

The ____________________ the surface, the less the friction is. The rougher the surface, the ____________________ the friction is.

2 Study the situations below. For each pair decide which would be easier to push and then circle A or B. Write why you think it would be easier to push.

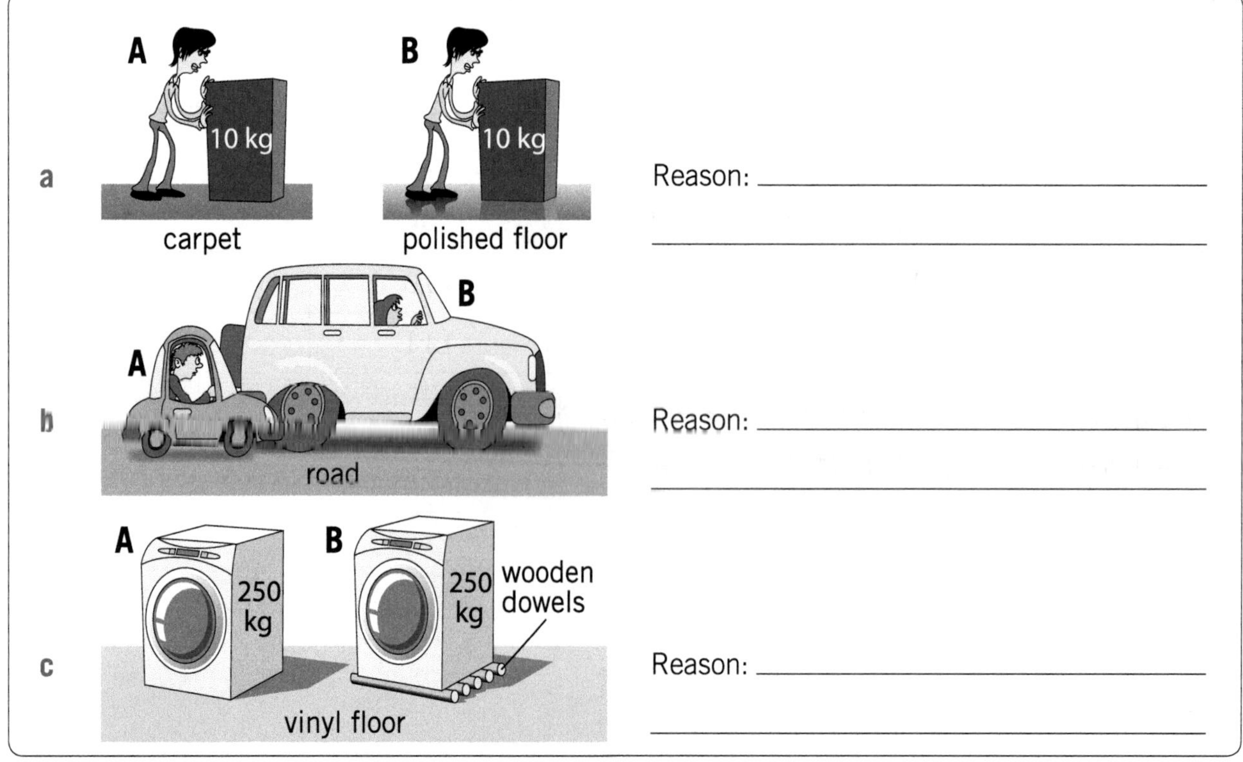

Reason: ____

____

Reason: ____

____

Reason: ____

____

Exercise 5 Applying information

After reading Reducing friction on page 63

1 Think about why you slip on a wet floor, why oil is used in a car engine, why you use grease on a bicycle chain, why you put soap on your finger to remove a tight ring, and why your mouth produces saliva.

Water, oil, grease, soap and saliva are all substances that ______________________________.

These substances are called ______________________________.

2 Lindy and Julie are having a race on a dry water slide. Lindy is sliding on a smooth mat and Julie is using roller skates.

Who do you think will reach the bottom first? Why? Explain your answer in terms of reducing friction.

____

____

____

3 Go back to your solutions for Problem 1 (moving the stone block) on page 18 of this workbook. Did you think of reducing the friction? If not, can you think of a way of moving the stone block by reducing the friction?

____

____

4 Go back to your solution for Problem 2 (stopping Bozo from killing himself) on page 18 of this workbook. He will only be stopped in his fall if the force to stop him balances the force pulling him downward.

a What is the force pulling him downward?

____

b What is the force you have suggested to stop him?

____

____

Exercise 6 Interpreting illustrations

The illustrations in a textbook are just as important as the text itself—sometimes more important. They may be drawings, cartoons or photos.

1 Look at the cartoon on page 62.

a What do the lines represent on the left hand side of the smaller bookcase?

____

b In which two ways does the cartoon indicate that a large force is needed to move the large bookcase?

____

2 Look at the drawing of a skateboard on page 63.

a What does the circular part on the left of the diagram represent?

____

b What do the ball-shaped things in the diagram do?

____

c Do a sketch in the box below to show how the ancient Egyptians were able to move the huge stone blocks needed to make the pyramids, by putting logs under them to reduce friction.
Write a caption for your sketch.

Your sketch goes here

Caption ____

____

3 Look at **Fig 2** on page 63 which shows a human hip.

a What does the blue part of the diagram represent?

____

b What is the pink part?

____

c Using the diagram, explain in your own words how your hip works when you walk.

____

____

____

d Find the reference to **Fig 2** on page 63 and use it to fill in the missing words.

Where ______________ slide over each other at ______________ it is important to reduce ______________. Because of this there is a ______________ fluid between them to make them ______________ more easily.

Exercise 7 Explaining in simple terms

After reading Gravitational forces on page 67

1 Suppose you are talking to a group of 8-year-olds about **gravity**. How would you explain it to them? (Give examples and use language that they would understand.)

2 Suppose you are talking to a group of 8-year-olds about **mass** and **weight**. They think **mass** and **weight** are the same. How would you explain these terms to them so that they understand the difference? (Give examples and use language that they would understand.)

Roundup

Exercise 8 Completing a mind map

By reading the textbook and doing the workbook exercises, you have a lot of information in your brain.

You need to organise the information into groups and subgroups to show how all the ideas link together. When you put these groups and their links together, you have a mind map.

Here is the author's mind map—with some information missing for you to complete by filling in the blank boxes.

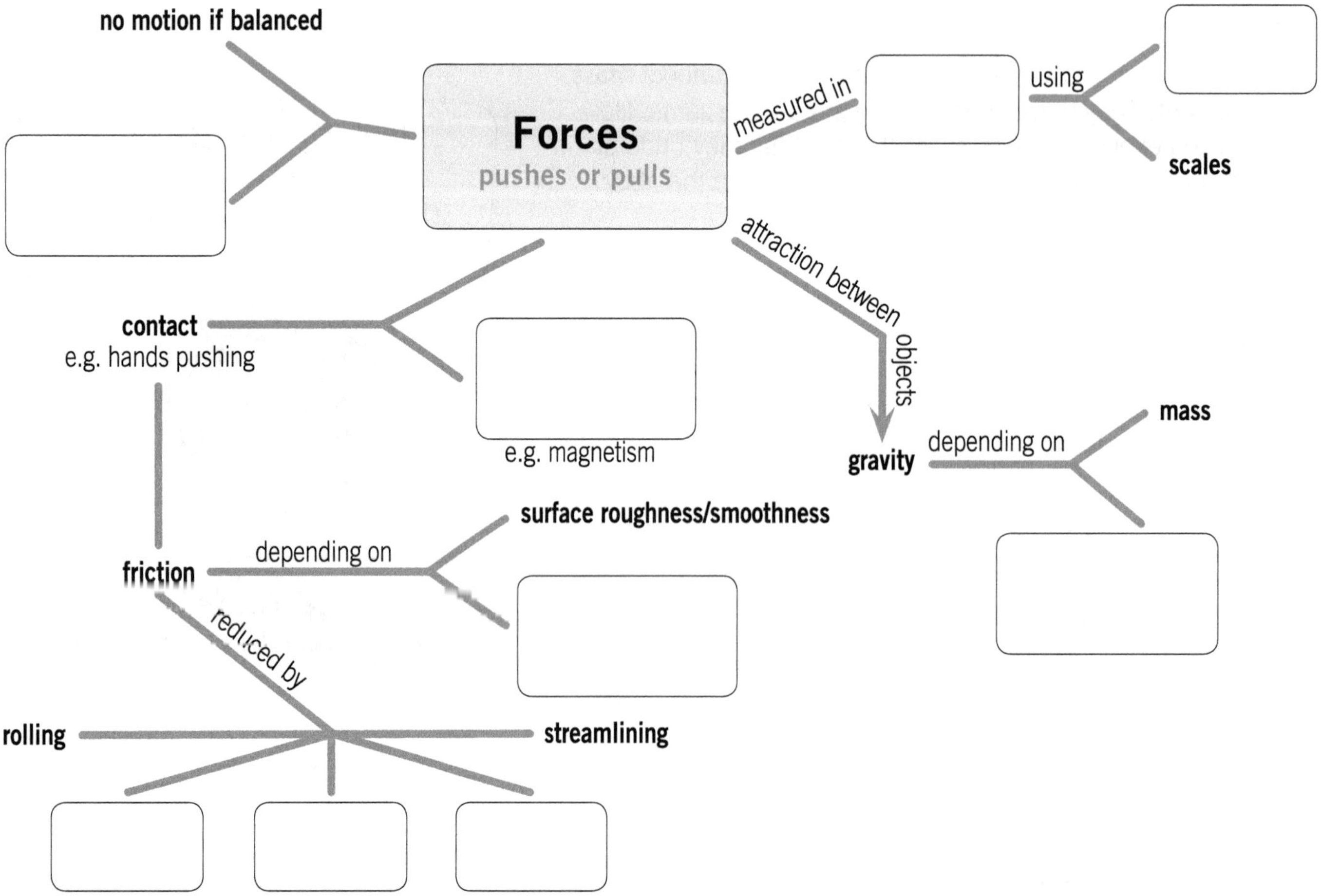

Forces
pushes or pulls

no motion if balanced

measured in

using

scales

attraction between objects

gravity

depending on

mass

contact
e.g. hands pushing

e.g. magnetism

friction

depending on

surface roughness/smoothness

reduced by

rolling

streamlining

Chapter 4

The living world

Overview

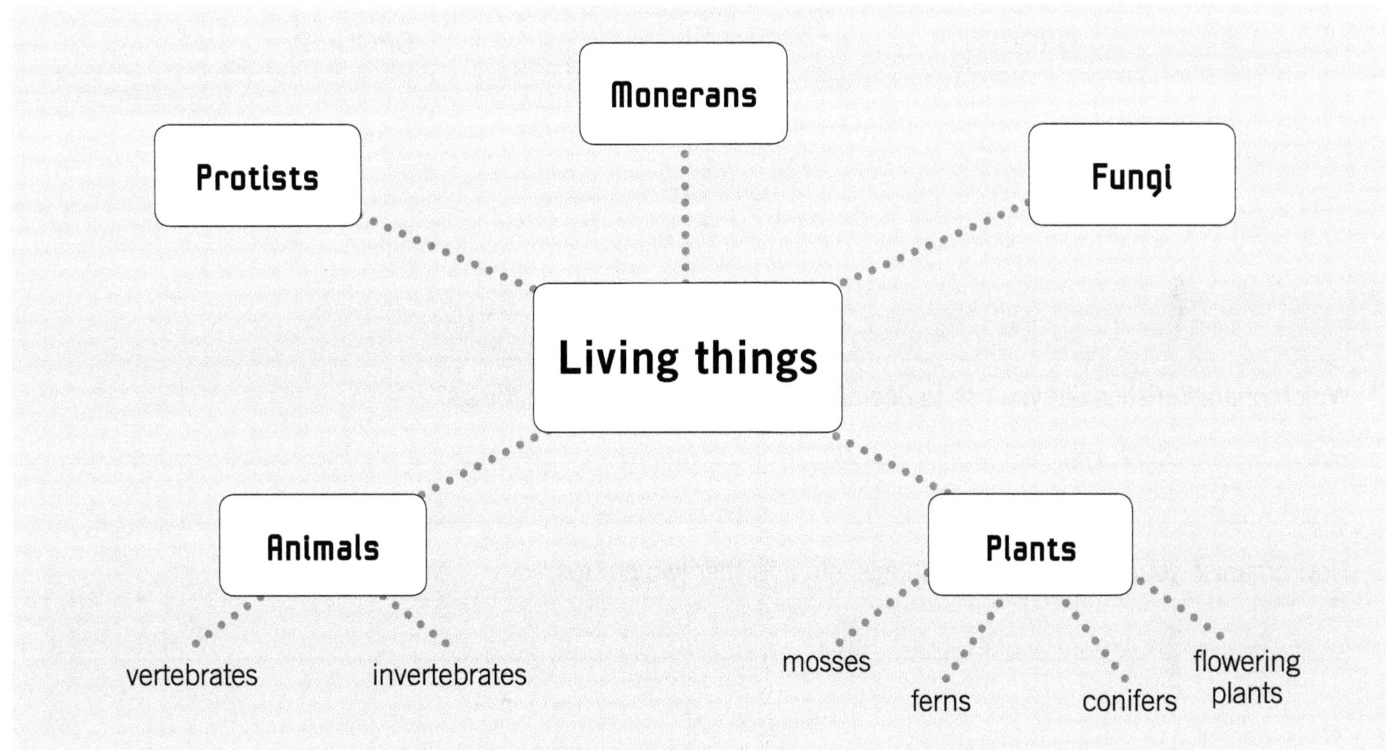

What do you know already?

Exercise 1 Grouping

With your teacher, go to an open space and make a single circle. When your teacher calls one of the following groups, the members with that characteristic are to make a smaller group inside the circle. (Your teacher may add other characteristics and will call them in any order.)

girls	students over 150 cm tall	girls who like science
boys	students whose surname starts with M	boys who play cricket
students with a backbone	students with blonde hair and blue eyes	students who can whistle
students with black hair	boys born in winter (June–August)	students born overseas

Exercise 2 Classifying

In small groups, complete the following activities.
Here is a list of living things.

dolphin	tree	flower	elephant	snail	toad
mushroom	seaweed	crab	jellyfish	fern	emu
goanna	stonefish	cactus	Venus fly trap	whale	plankton

1 Organise them into two groups.

GROUP 1	GROUP 2

2 Which characteristics did you use to decide how to group these living things?

____

____

3 Now organise your GROUP 1 living things into a further two groups.

4 Which characteristics did you use to decide how to group them this time?

____

____

5 Now organise your GROUP 2 living things into a further two groups.

6 Which characteristics did you use to decide how to group them this time?

____

____

During reading

Exercise 3 Classifying

After reading Classifying things on pages 77–78

Here is a list of objects.

beaker	tennis ball	relay baton
flute	test tube	basketball

1 In small groups, classify these objects into two groups based on the characteristic of shape.

Spherical shapes	Cylindrical shapes

2 Now classify the objects into three groups, based on the characteristic of function.

Musical instrument	Laboratory equipment	Sporting equipment

Exercise 4 Using a key

After reading Using keys on page 78

Study the following key for classifying students in a class.

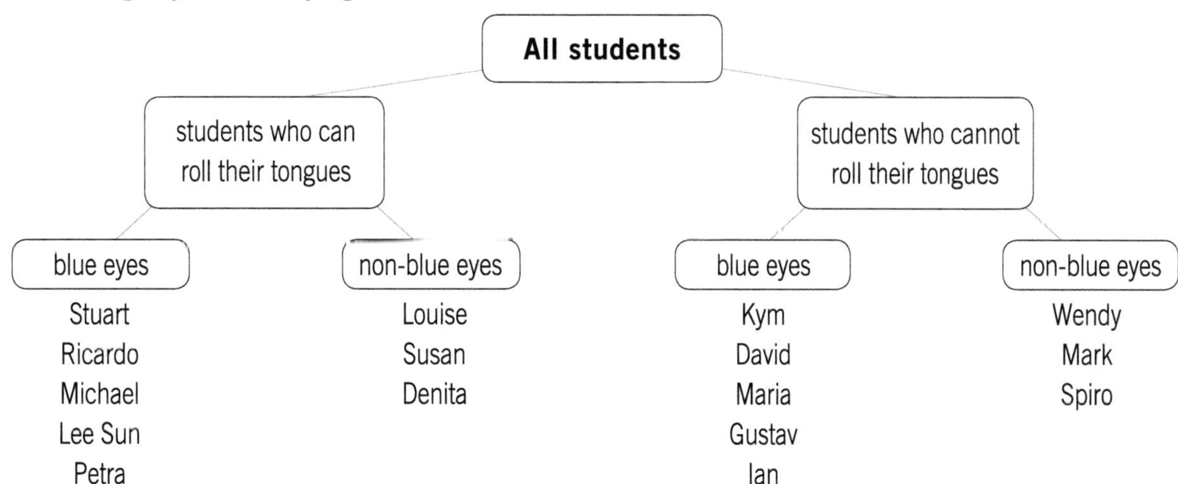

1 Name the students who can roll their tongues but do not have blue eyes.

____

2 What do Louise and Stuart have in common? ______________________________

3 How many students cannot roll their tongues and do not have blue eyes? ______________________________

4 Add your name to the group that applies to you.

Exercise 5 Making and justifying decisions

After reading Classifying living things on pages 80–81

Using the key on page 81, decide whether the following statements are true (T), false (F), or not enough information has been given to decide (?).

Circle your decision, making sure that you have a reason for each choice.

1	Fish have a changing body temperature.	T	F	?
2	Worms have a backbone.	T	F	?
3	Snails have a constant body temperature.	T	F	?
4	Reptiles breathe by gills.	T	F	?
5	Molluscs have a shell without a jointed covering.	T	F	?
6	Mammals have a constant body temperature and no feathers.	T	F	?
7	Worms have moist skin.	T	F	?
8	Arthropods breathe by lungs or air tubes.	T	F	?

Exercise 6 Comparing and contrasting

After studying the key on page 81

1 Which three characteristics do REPTILES and AMPHIBIANS share?

____

You can write about these **similarities** using the *comparison linking words* on page 92 of this workbook. Both reptiles and amphibians are vertebrates that have a changing body temperature and breathe by lungs or air-tubes.

2 What are the **differences** between REPTILES and AMPHIBIANS?

____

You can write about these differences using the *contrast linking words* on page 92 of this workbook. Reptiles have scaly skin, however amphibians have moist skin.

3 Use *comparison linking words* to write about the **similarities** shared by:

a MAMMALS and BIRDS ______________________________

b ARTHROPODS and WORMS ______________________________

4 Use *contrast linking words* to write about the **differences** between:

a MAMMALS and BIRDS

____

b ARTHROPODS and WORMS

____

Exercise 7 Completing an overview

After reading The five kingdoms on page 84

Complete the overview to show how biologists have classified all living things.

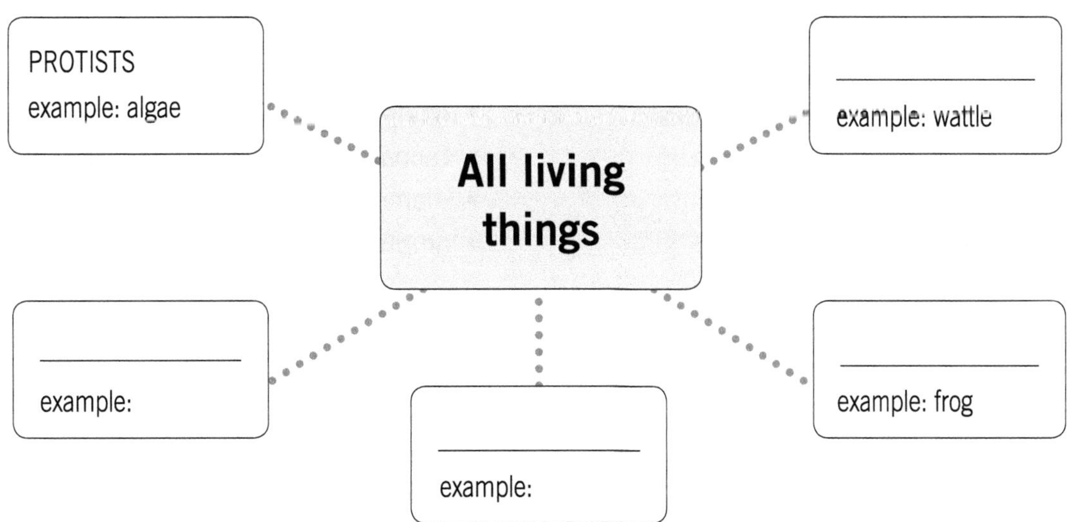

Exercise 8 Completing a table

After reading pages 85–88

Complete the table below. One column is done for you.

	Animal kingdom	Plant kingdom	Fungi kingdom	Protist kingdom	Monera kingdom
What are the characteristics of this kingdom?				simple structure (no roots, no stems, no leaves) usually unicellular most live in water contain chlorophyll and make own food	
What are some examples of this kingdom?				algae microscopic organisms	

Exercise 9 Completing a structured overview

After reading The animal kingdom on pages 91–92

Use the words listed below to complete the boxed overview.

- arthropods
- turtle
- spider
- fish
- small internal shell
- arachnids
- molluscs
- pelican
- prawn
- ant
- reptiles
- octopus
- human
- mammals
- amphibians
- snail
- salmon
- frog
- birds

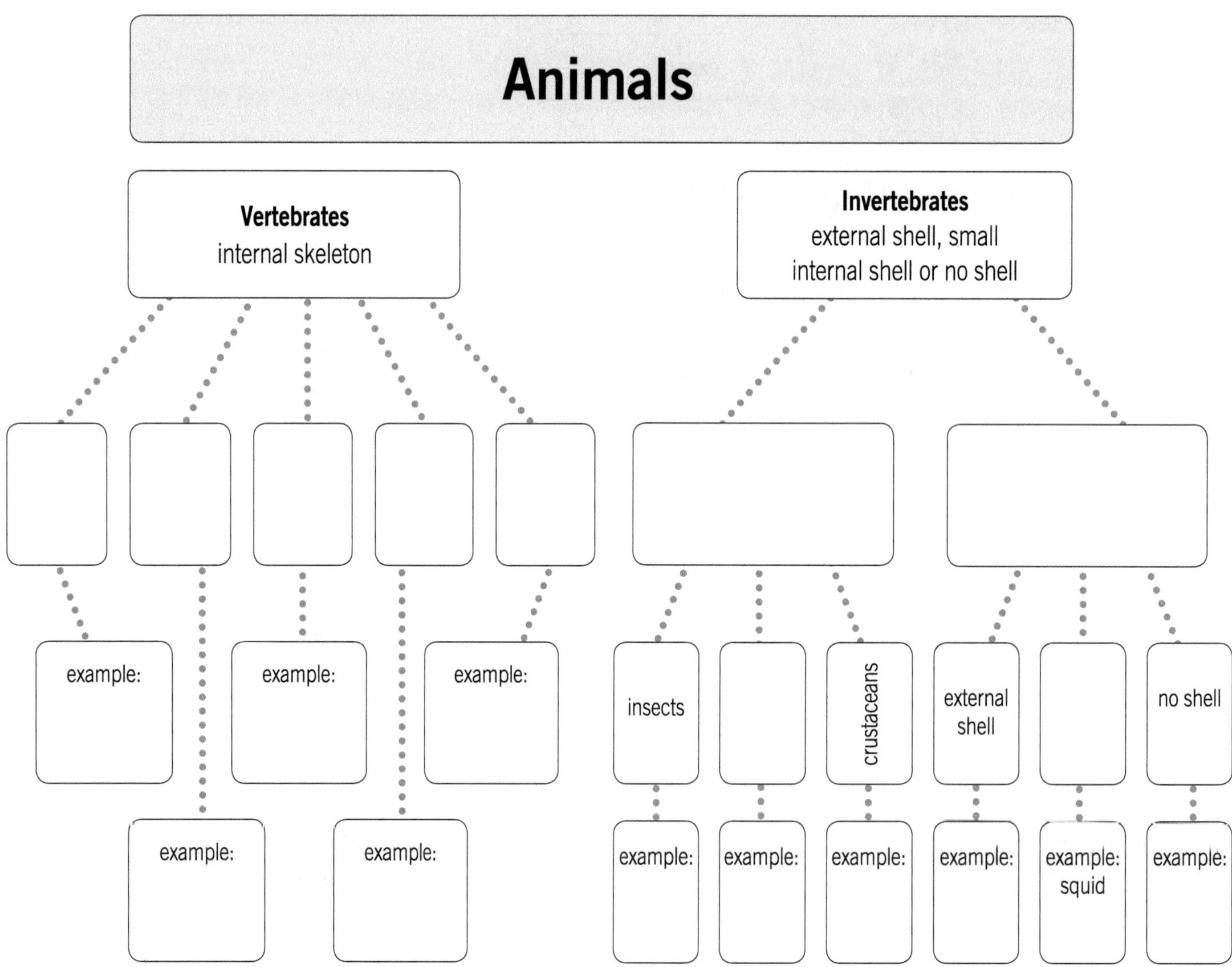

Exercise 10 Defining

After reading The plant kingdom on pages 95–96

In Science, you will be asked to define words. This language skill has its own structure and is different from describing or explaining. It is easy to define when you know the structure for expressing your knowledge. Here are some definitions. See if you can see a common pattern for writing a definition.

An astronomer is a scientist who studies objects in space.

A snake is an animal that has a changing body temperature, scales and no legs. An example is a python.

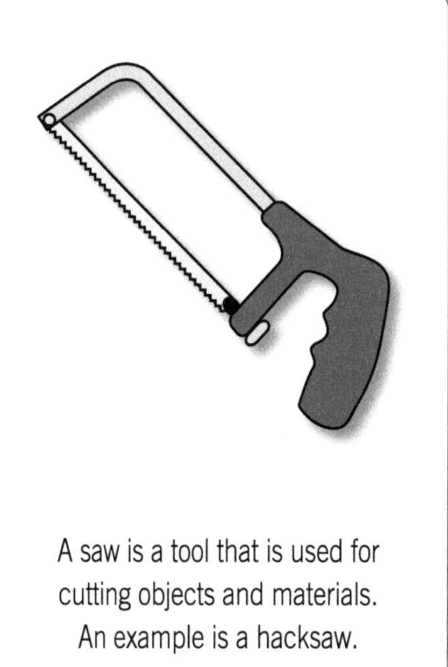

A saw is a tool that is used for cutting objects and materials. An example is a hacksaw.

A definition is a group of words which has the following pattern.

A **(word to be defined)** is a **(larger group to which it belongs)** that / which / who **(list the characteristics that make the word different from the rest of the group)**.

An example is ______________________ .
(An example may not need to be added.)

Let's define the word **bicycle**.

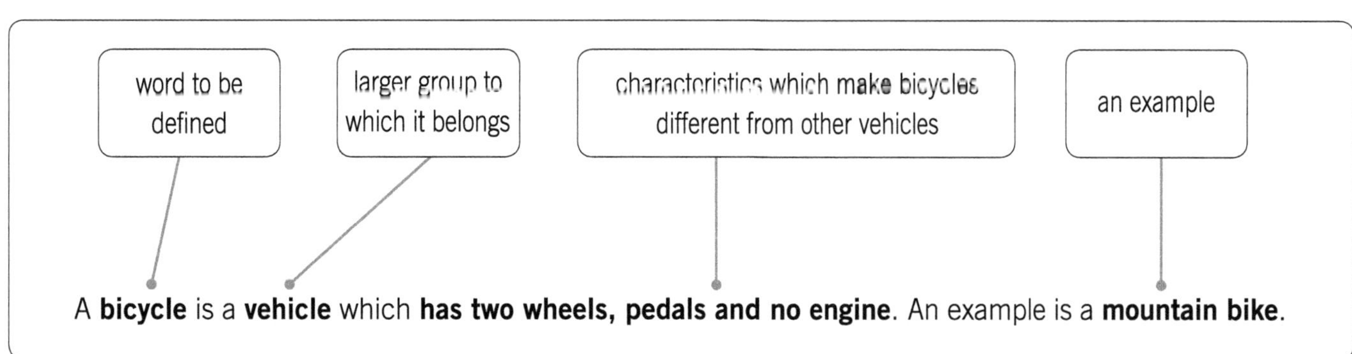

Here is an example of how to use the plant key on page 95 to define the word **fern**.

A fern is a plant which has a stem and makes spores, not seeds.

Using the example above as a model and the key on page 95, complete the following definitions.

1 Conifers are plants that

____

____

2 Flowering plants are plants that

____

____

3 Mosses are plants that

____

____

Roundup

Exercise 11 Classifying

Imagine you are a biologist. Work in a group to decide how you would classify the following organisms.

An organism that has no backbone, no jointed covering and a shell. KINGDOM: ____________ Type of organism: ____________	An organism that is multicellular, has no stems, no leaves, no roots and contains chlorophyll. KINGDOM: ____________ Type of organism: ____________
An organism that has a stem, reproduces by seeds, but has no flowers. KINGDOM: ____________ Type of organism: ____________	An organism that has a backbone, a constant body temperature and feathers. KINGDOM: ____________ Type of organism: ____________

Exercise 12 Writing a paragraph

Suppose you were given the following task.

> Write a paragraph to compare and contrast fish and amphibians.

When you are given a task like this, you should follow the steps in this reading and writing roadmap.

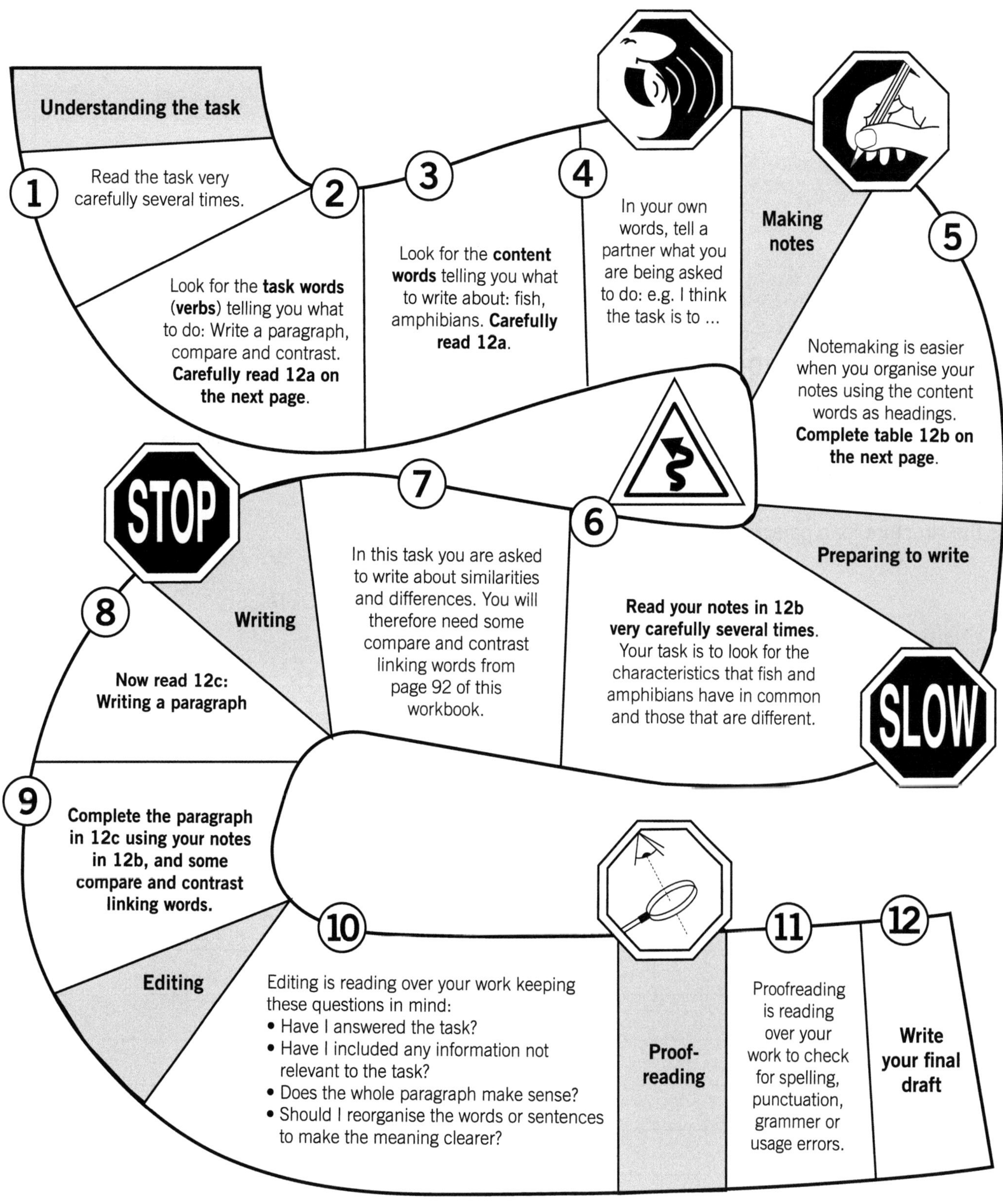

12a Task words and content words

Task words are words (verbs) which tell you what to do, e.g. explain, write, discuss, compare, contrast, research, etc. **Content words** tell you what to write (or speak) about. In this paragraph, you have to write about fish and amphibians.

12b Notemaking

Fish	Amphibians

12c Writing a paragraph

A paragraph is a 'parcel' of information, usually containing several sentences which describe or develop one main idea. Paragraphs usually begin with a topic sentence which states the main idea of the paragraph. (This main idea is often a generalisation.) The remaining sentences in the paragraph then give details supporting the topic sentence.

This is the structure for a paragraph.

TOPIC SENTENCE + SENTENCES WITH SUPPORTING DETAILS

Complete the task remembering to use *compare and contrast linking words.*

TOPIC SENTENCE

Fish and amphibians are both members of the animal kingdom. They have many similarities and some differences. ______________________________

SUPPORTING SENTENCES

____
____
____
____
____
____
____
____
____

Chapter 5

Earth in space

Overview

Cause	Effect
night / day / ROTATION / axis / sunlight	**Day and night**
revolution / sun / axis tilted / window in Southern Hemisphere	**Seasons**
Earth / moon / sunlight	**Phases of the moon**
solar eclipse: sun, moon, Earth; lunar eclipse: sun, Earth, moon	**Eclipses**
low tide / high tide / high tide / low tide / moon	**Tides**

What do you know already?

Exercise 1 Organising prior knowledge

In this chapter, you will be learning about the sun, Earth and moon. You already know some facts about these three bodies in our solar system. Fill in the table below using your own knowledge and questions before reading the textbook. Try to think of at least three responses for each body.

What do you...	SUN	EARTH	MOON
know you know about the ... ?			
think you know about the ... ?			
think you'll learn about the ... ?			

Keep your answers in mind as you read the chapter.

Exercise 2 Describing

Look at the photo of the Earth in space on page 102. Imagine you are standing on the moon and you are asked to describe what you are seeing. Complete the description below.

From where I am standing, the Earth looks ______________________________

____

____

____

____

During reading

Exercise 3 Recalling information

After reading How the Earth moves on pages 104–105

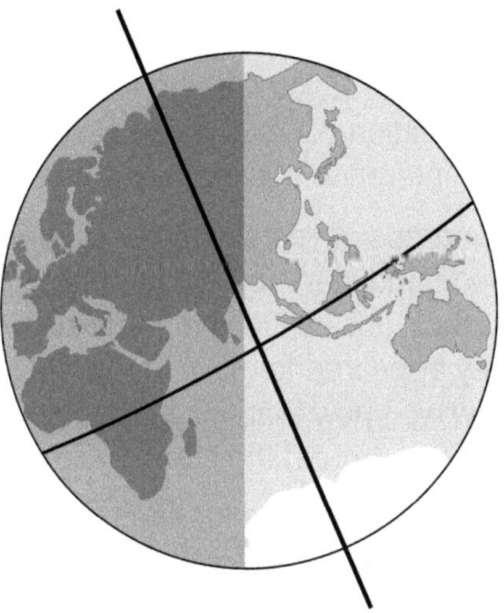

1 Put the following labels in the correct places on the diagram. Do not refer to your textbook.

- axis and equator
- North Pole and South Pole
- east and west
- side in darkness
- Northern Hemisphere and Southern Hemisphere
- side in sunlight

2 Use an arrow to show the direction in which the Earth rotates.

3 The five definitions below are mixed up. Rewrite them correctly.

a An orbit is / the spinning of the Earth on its axis.

____

b The equator is / the imaginary line through the centre of the Earth from pole to pole.

____

c An axis is / the movement of the Earth around the sun.

____

d Rotation is / the path the Earth follows around the sun.

____

e Revolution of the Earth is / an imaginary line dividing the Earth into two hemispheres.

____

Exercise 4 Interpreting diagrams

After reading What causes the seasons? on page 107 and doing Investigation 12 on pages 108–109

1 Examine **Fig 4** on page 107 carefully. Why is it warmer at the end of beam A than at the end of beam B?

____

2 Examine the diagram at the bottom of page 107 carefully.

a In the Southern Hemisphere, which three months have the longest hours of sunlight?

____

b In the Northern Hemisphere, which three months have the longest hours of sunlight?

____

c In the Northern Hemisphere, what is the season as the South Pole starts to tilt away from the sun?

____

d If the Earth's axis did not tilt, predict what would happen to the seasons.

____

Exercise 5 Making inferences

After reading Studying lunar craters on page 113

In Chapter 2 you learned that an **inference** is a possible explanation you can make based on your observations. The inference may or may not be true.

Hundreds of years ago, people **inferred** that the sun, moon and stars revolved around the Earth. They based this on their observations of these heavenly bodies as they appeared to move across the sky from east to west. Now we know that this inference was wrong. The Earth revolves around the sun.

Make an inference for each of the following. Check the inferences made by other students in your class.

1 **Observation**: It is 12 noon when you wake up in a strange place. You look out the window and it is very dark. There is no sunlight.

Inference: ______________________________

____

____

____

2 **Observation**: You look up one night to find the moon but it is not visible in any part of the sky.

Inference: ______________________________

____

____

____

3 **Observation**: The moon has many craters on its surface.

Inference: ______________________________

____

____

____

Exercise 6 Illustrating

After reading Phases of the moon on page 116

Illustrate the phase of the moon in the following boxes. The first one is done for you.

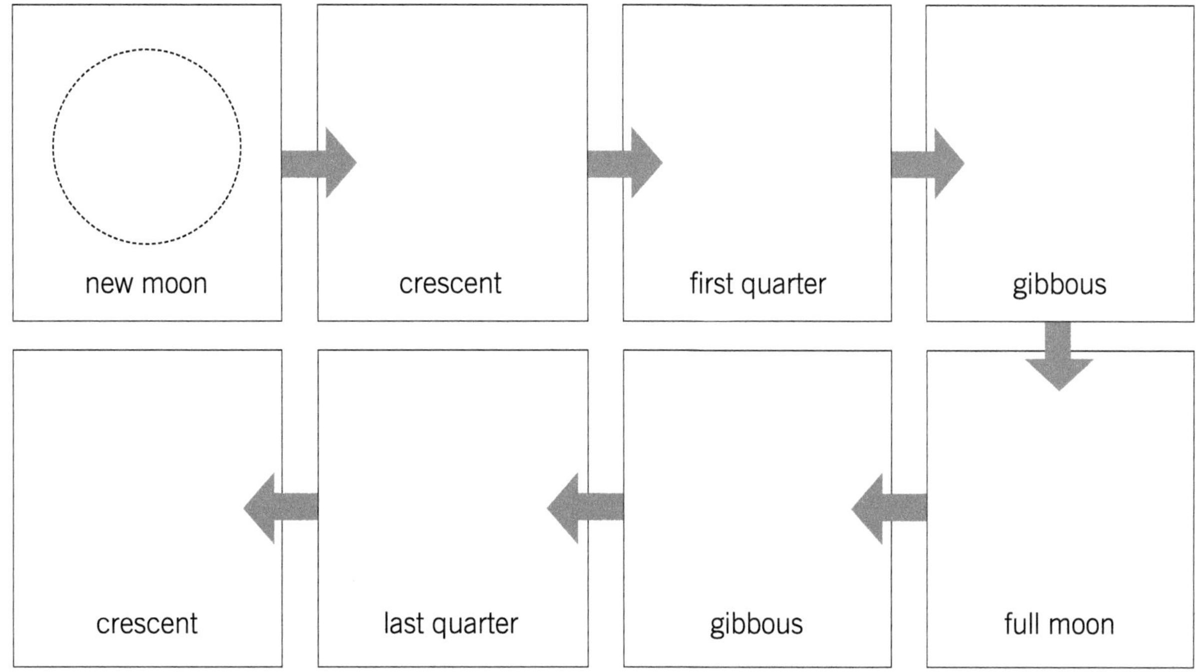

Exercise 7 Completing sentences and illustrations

After reading Solar eclipses and Lunar eclipses on pages 117 and 118

1 Complete these sentences in your own words.

A solar eclipse occurs when ______________________________

____

A lunar eclipse occurs when ______________________________

____

2 Imagine you are on Earth looking indirectly (and safely) at the sun during a total solar eclipse. Draw a series of sketches to illustrate what you would see from Earth.

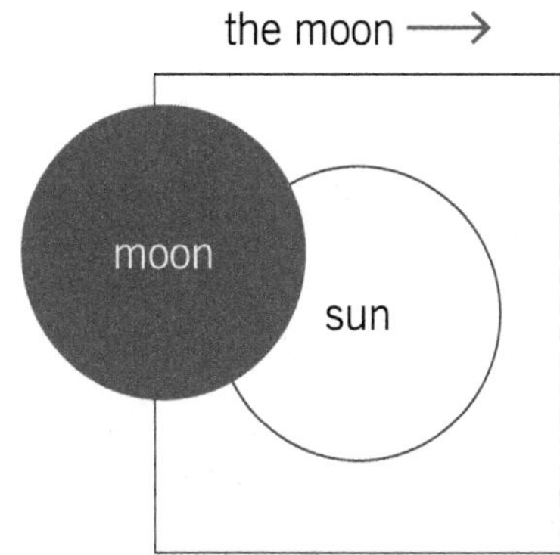

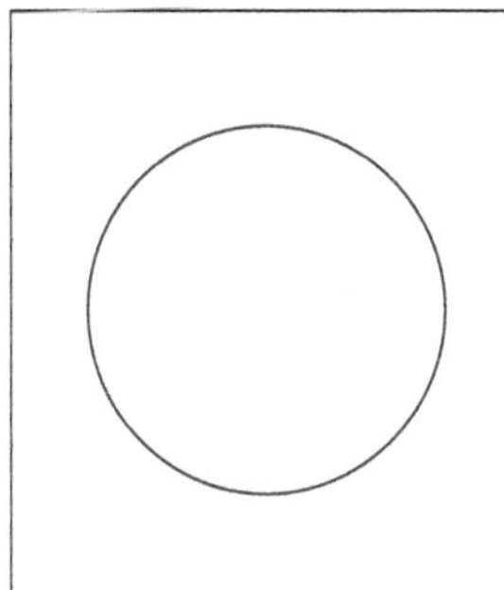

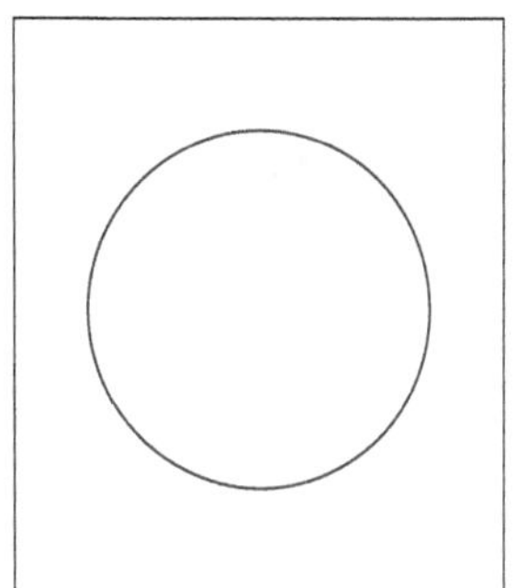

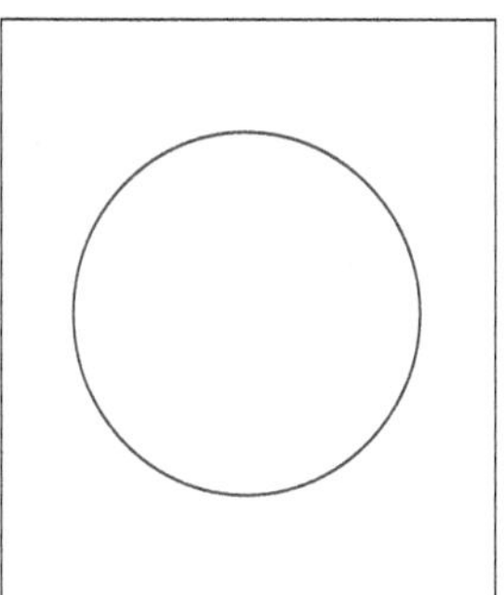

Exercise 8 Completing cartoons

After reading Tides on page 119

Complete the cartoons below.

Exercise 9 Using cause and effect linking words

Throughout this chapter, there are many examples of cause and effect.

1 Complete the table by adding the missing causes or effects. The first one is done for you.

Cause	Effect
a Rotation of the Earth on its axis.	day and night
b ______________________	seasons
c Gravitational force of attraction between the Earth and the sun.	______________
d The moon revolves around the sun in 29.5 days and it rotates on its axis every 29.5 days.	______________
e The same area of the moon's surface is lit by the sun, but different amounts of its surface can actually be seen from Earth.	______________
f Earth's spinning on its axis and the gravitational pull of the moon (and to a lesser extent, the sun).	______________
g ______________________	lunar eclipse

2 You can put the actions in the table above into cause and effect sentences by using *cause and effect linking words*. You can use some of the words if you put the cause first, and other words if you put the effect first, e.g. the first one can be written in two ways.

a The rotation of the Earth on its axis results in day and night. **OR**

Day and night occurs as a result of the rotation of the Earth on its axis.

Use some of the *cause and effect linking words* on page 92 of this workbook, to link in sentences the remaining causes and effects in the table above.

b ____

c ____

d ____

e ____

f ____

g ____

Exercise 10 Using role-play

Organise yourselves into groups of four.

Here are your roles.

Person 1 The Earth

Person 2 The moon

Person 3 The sun

Person 4 The narrator

Your audience: A group of 6-year-olds

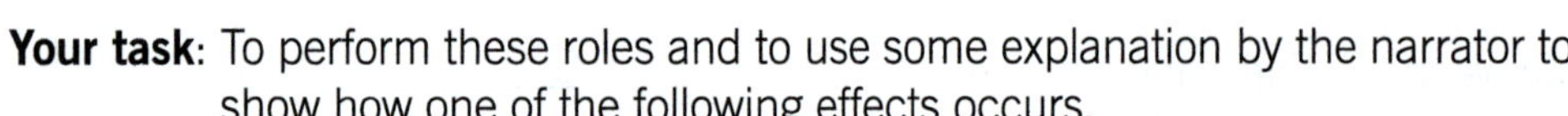

Your task: To perform these roles and to use some explanation by the narrator to show how one of the following effects occurs.

Step 1 Decide who will play each part.

- Revolution of the Earth around the sun, and the moon around the Earth
- Day and night
- The seasons
- Phases of the moon
- Solar and lunar eclipses
- Tides

Step 2 Organise how you will present the information through movement and an explanation by the narrator.

Step 3 The narrator will introduce the role-play by saying something like this:

"Hello, girls and boys. Have you ever wondered how the Earth and moon move around in space? Today we are going to try to show you the movements that take place by acting it out for you. Janet will be the Earth, Sonia will be the moon, and Wesley will be the sun. My name is Michael and I will do most of the talking."

Step 4 Rehearse until you are sure the performance is entertaining, simple and appropriate for 6-year-olds to understand and enjoy.

Step 5 Why not organise an audience of 6-year-olds to entertain? Who knows? They might just learn something!

Exercise 11 Reviewing

Turn back to your responses to Exercise 1 on page 36 of this workbook. After reading through the chapter in the textbook and completing the Exercises in this workbook, you can now check your responses.

- Are all your KNOW YOU KNOW facts correct? If not, correct them now.
- Are all your THINK YOU KNOW facts correct? If not, correct them now.
- Did the textbook include your THINK YOU'LL LEARN responses? If not, look them up in another book or ask your teacher.

Chapter 6
Separating mixtures

Overview

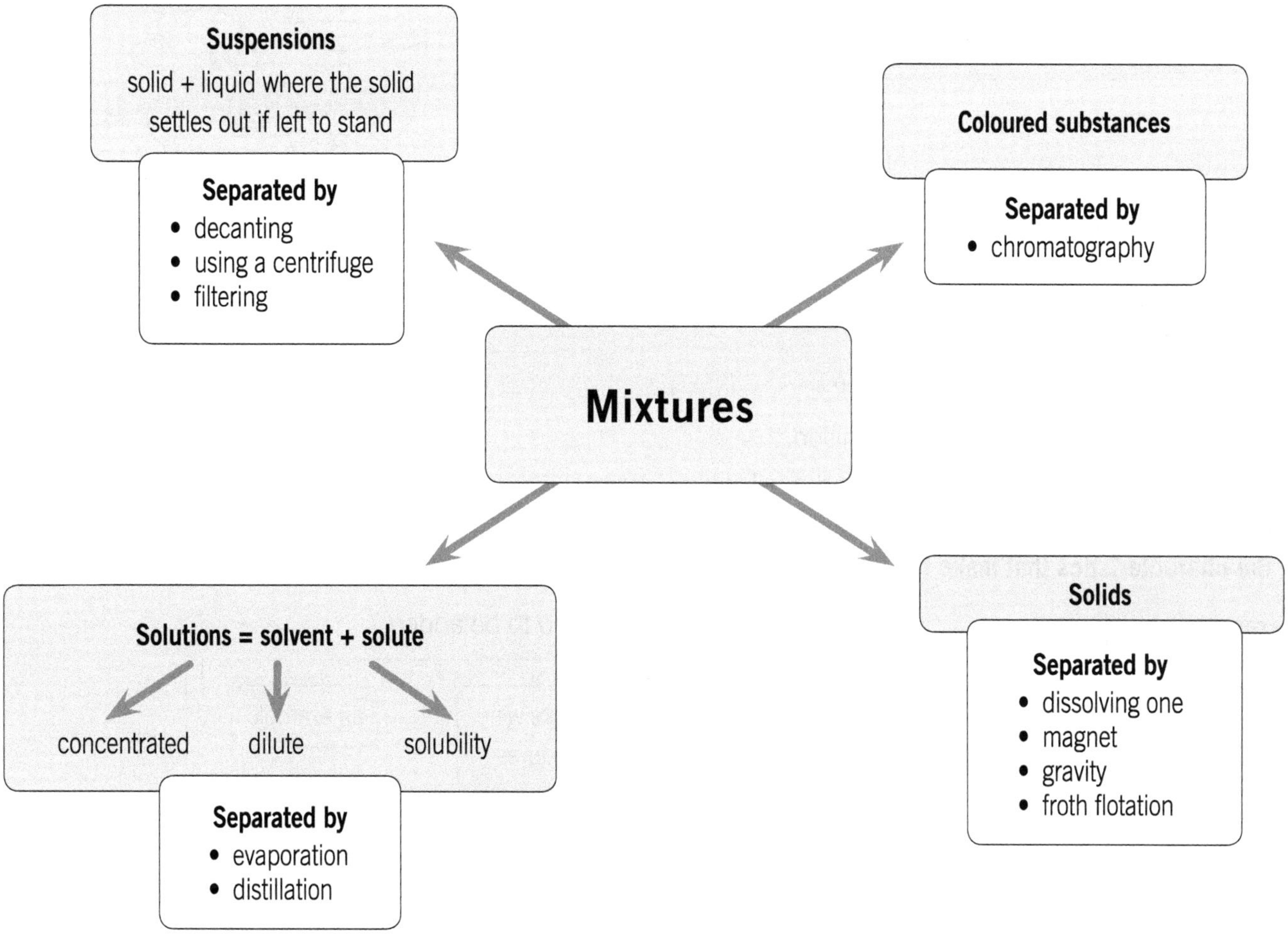

Mixtures

Suspensions
solid + liquid where the solid settles out if left to stand

Separated by
- decanting
- using a centrifuge
- filtering

Coloured substances

Separated by
- chromatography

Solutions = solvent + solute
concentrated
dilute
solubility

Separated by
- evaporation
- distillation

Solids

Separated by
- dissolving one
- magnet
- gravity
- froth flotation

What do you know already?

Exercise 1 Problem-solving

Problem: Imagine you are a farmer and there is a severe drought. Your sheep desperately need water, and the only water is bore water which is too salty and muddy.

How can you produce fresh water from the bore water?

Suggest a possible solution on the next page.

Suggest a possible solution.

____

____

____

____

____

____

____

Yeech! No wonder the sheep don't drink this stuff!

During reading

Exercise 2 Defining

After reading What's a mixture? on page 126

Here is a reminder of how to write a definition.

A definition is a group of words which has the following pattern:

A **(word to be defined)** is a **(larger group to which it belongs)** that / which / who **(list the characteristics that make the word different from the rest of the group)**.

An example is ____________ . (An example may not need to be added.)

word to be defined | larger group to which it belongs | characteristics which make wine different from other mixtures | an example

Wine is a **mixture** which contains **alcohol** in **grape juice**. An example is **champagne**.

Use this pattern to complete the definitions below.

1 A **mixture** is a substance which ____

____

2 **Pure substances** are substances that ____

____

3 **Lipstick** is a substance which ____

____

4 **Smoke** is a mixture that ____

____

Exercise 3 Writing definitions

After reading Solutions on page 127

Complete these definitions.

1 A **solvent** is a substance which ____________________

An example is ____________________

2 A **solute** is a substance which ____________________

An example is ____________________

3 **Suspension** is a mixture that ____________________

An example is ____________________

4 A **colloid** is a mixture which ____________________

An example is ____________________

5 An **emulsion** is a mixture which ____________________

An example is ____________________

Exercise 4 Observing

After doing Investigation 13 on pages 128–129

1 List four substances in your home that are soluble.

____

2 List four substances in your home that are insoluble.

____

3 To test the solubility of various kitchen substances, Megan measured how much of each substance dissolved in 100 mL of water. Here are her results.

a Which substance is most soluble in water? ______________________________

b Which substance is least soluble in water? ______________________________

c From Megan's results, how much sugar would dissolve in a litre of water? ______ g

Substance	Amount (in grams) that dissolved in 100 mL water
salt	36
sugar	204
curry powder	1
baking soda	10

Exercise 5 Explaining in your own words

After reading Solubility on page 129

1 Complete the cartoon.

2 In Exercise 4, Megan could dissolve only 36 grams of salt in 100 mL of water. Describe what she could do if she wanted to dissolve more salt in 100 mL of water.

____

____

Exercise 6 Completing a table

After reading Separating suspensions on pages 133–134

Complete the table below.

Method of separating	Describe what happens in this process	Everyday example of this type of separation
	The liquid is poured off while keeping the solid in the container.	
CENTRIFUGE		
		vacuum cleaner

Exercise 7 Completing a table

After reading Separating solutions on page 136

Complete the table below.

Method of separating	Describe what happens in this process	Everyday example of this type of separation
	The liquid is heated so that it turns into a vapour and goes into the air, leaving the solid behind.	
	Water and alcohol have different boiling points so they evaporate at different temperatures.	

Exercise 8 Completing a table

After reading Separating solids on page 139

Complete the table below.

Method of separating	Describe what happens in this process	Everyday example of this type of separation
DISSOLVE ONE THEN FILTER		separating salt from sand
	The magnet separates the magnetic minerals from other minerals.	
FROTH FLOTATION		gold panning

Exercise 9 Defining and interpreting a diagram

After reading Separating colours on page 141

1 Complete the definition.

Chromatography is a process in which

____

____

____

An example is ______________________________

2 Examine the diagram on the right then answer the questions.

The diagram shows a filter paper with an ink spot after water has been dripped onto it.

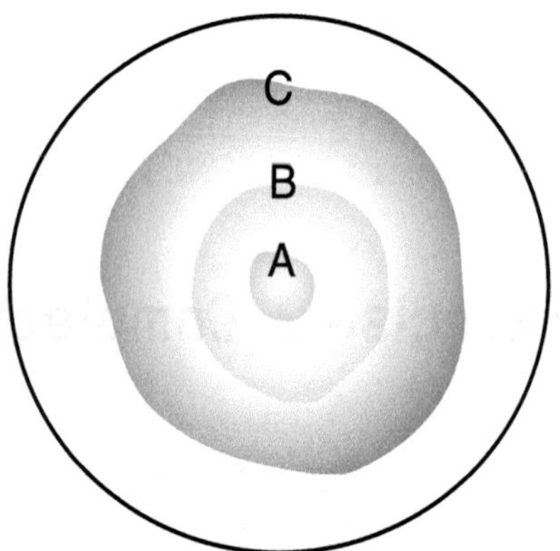

a Which ring (A or B or C) contains the most soluble colour? __________

b Which ring (A or B or C) contains the least soluble colour? __________

c How many colours does the ink contain? __________

Roundup

Exercise 10 Matching separating processes with mixtures

Write the best method for separating each of the following mixtures. Choose from the methods in the box on the right. You will need to use some processes more than once and not use some other listed processes.

1 salt from sea water ______________________

2 alcohol from water ______________________

3 chalk from water ______________________

4 water from clothes ______________________

5 cream from milk ______________________

6 steel cans from aluminium cans ______________________

7 dirt from air ______________________

8 blue ink from red ink ______________________

9 the substances in crude oil ______________________

filtering
using a centrifuge
evaporation
distillation
chromatography
using a magnet
froth flotation
gravity separation
decanting

Exercise 11 Reviewing your problem-solving

Go back to your solution for the problem on pages 43–44 of this workbook. Now that you know more about separating mixtures, could you improve your solution for separating the salt and mud from the water? How?

____

____

____

____

____

Exercise 12 Completing a flow chart

Ian and Luke drew the flow chart below to describe their solution to the *third problem* in the Getting started on page 125 of the textbook.

Inside the boxes are the names of mixtures or pure substances. Next to the arrows are the methods of separation used. In rushing to get lunch, they forgot to fill in one of the boxes. They also forgot to label three of the arrows. Complete the flow chart by filling in the box and labelling the arrows.

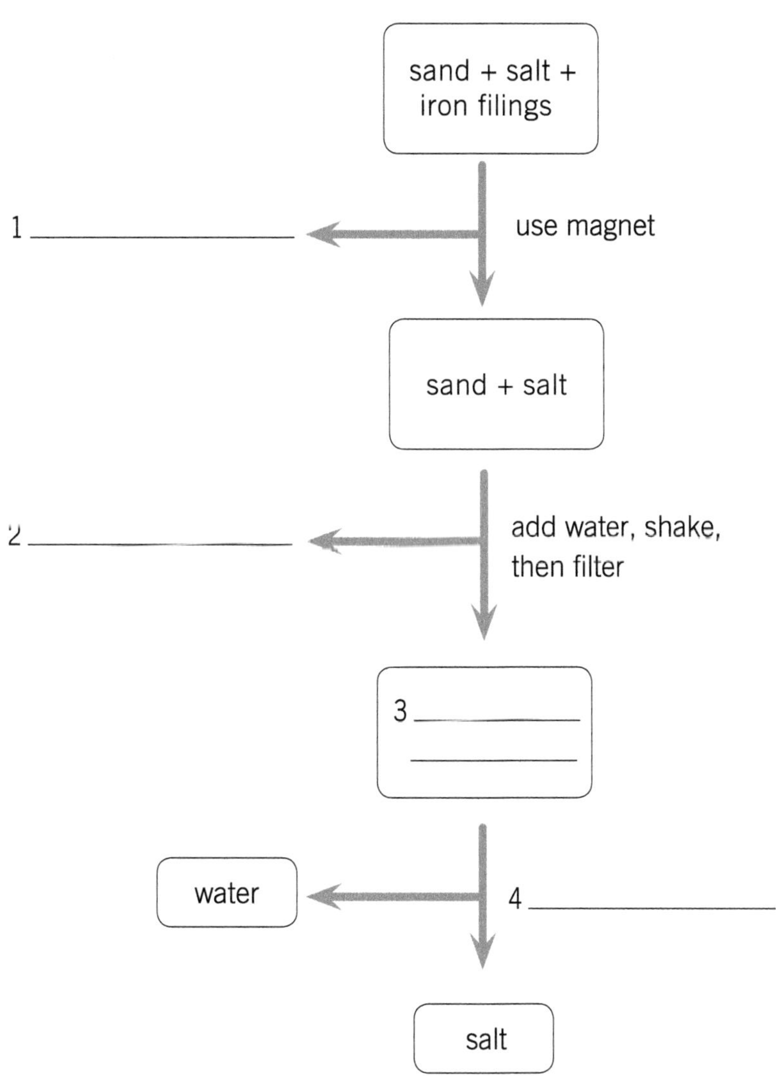

Exercise 13 Matching words and meanings

Match the words with their meanings, by writing the correct number on the line.

1 solution
2 evaporation
3 condensation
4 saturated
5 soluble
6 solute
7 solubility
8 solvent
9 suspension
10 mixture
11 decanting
12 insoluble
13 filtration
14 chromatography

___ Two or more substances mixed together

___ The process in which a liquid turns to vapour

___ Not able to be dissolved

___ The amount of a substance that can be dissolved in a liquid

___ A liquid in which substances dissolve

___ Separation process which works because some colours are absorbed more easily by paper than others

___ A mixture in which the tiny bits of substance spread evenly throughout a liquid are too small to be seen

___ No more solute can be dissolved

___ A substance that dissolves in a liquid

___ The process in which a vapour changes into a liquid

___ Separating a solid from a liquid using a filter funnel and filter paper

___ A mixture in which small bits of a substance are spread evenly throughout a liquid but are not dissolved

___ Gently pouring off a liquid, leaving the solid in the container

___ Able to be dissolved

Chapter 7
Living places

Overview

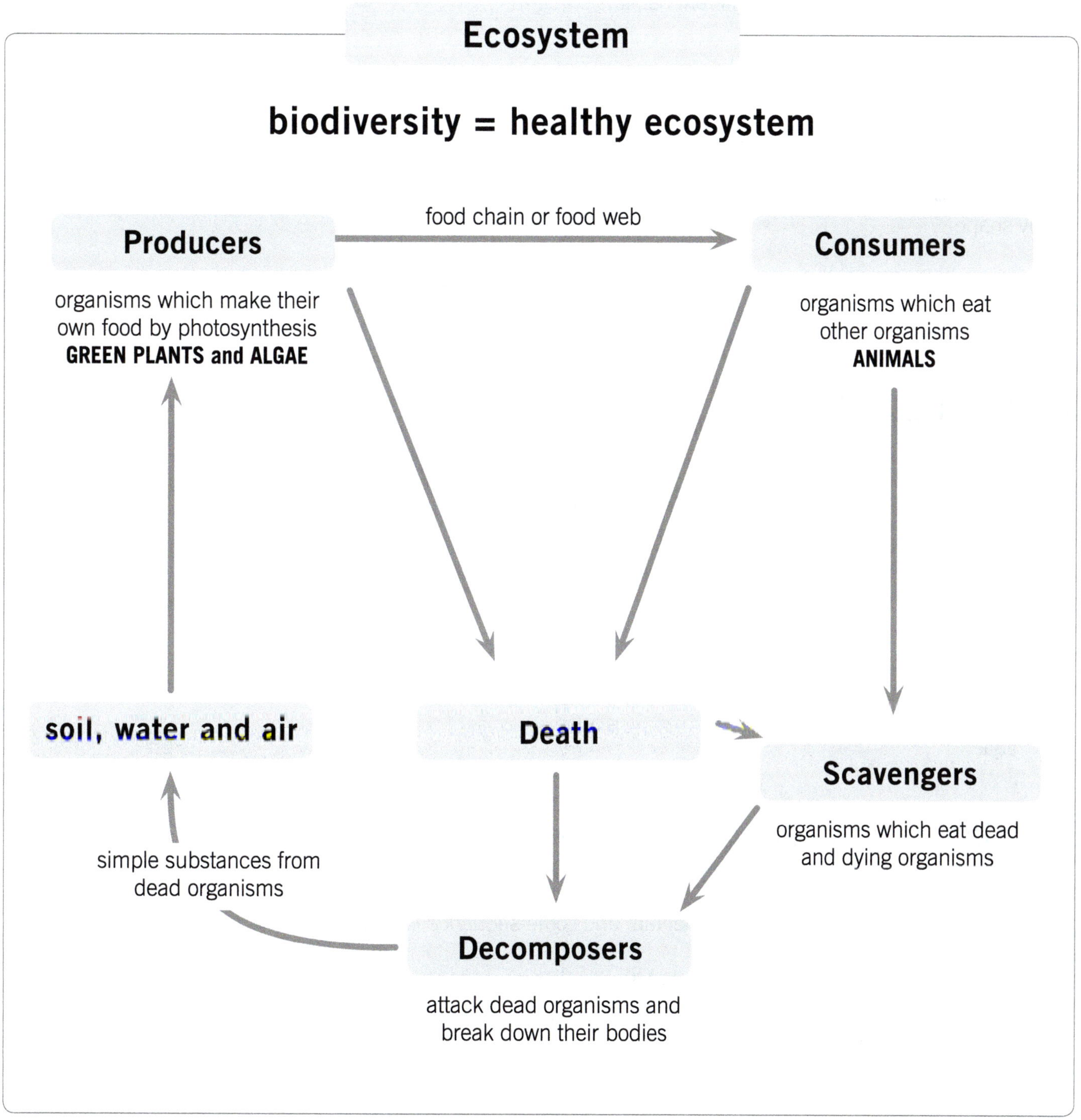

Ecosystem

biodiversity = healthy ecosystem

What do you know already?

Exercise 1 Selecting relevant information

Form small groups to discuss human survival.

Which characteristics do humans have that have enabled humans to survive in the many different habitats on Earth?

From the list of human characteristics below, select the information relevant to our survival in the many different habitats on Earth.

- hair colour
- skin colour
- sense of humour
- height
- body temperature
- body shape
- walking upright
- running speed
- intelligence
- language
- curly hair
- weight
- hairy arms and legs
- living in families
- long fingernails
- hand shape and structure
- living in communities
- eyesight

Tick your selections and share your decisions, and your reasons for doing so, with the rest of the class.

During reading

Exercise 2 Completing diagrams

After reading Food chains on page 150

Fill in the boxes to complete these food chains.

lettuce	is eaten → by		is eaten → by		is eaten → by	snake
algae	is eaten → by		is eaten → by		is eaten → by	shark

Exercise 3 Translating from diagram to text

After reading Photosynthesis and respiration on page 151

Study the leaf diagram carefully. Use the diagram and some *sequence linking words* from page 92 of this workbook
to complete this paragraph describing the photosynthesis process in your own words.

Photosynthesis is the process in which plants produce their own food. To begin with, ________________

____

____

____

Exercise 4 Translating from text to diagram

After reading Food webs on page 154

Draw a food web for the following situation using arrows and the standard structure used by biologists for drawing food webs.

> In my garden I observe birds, bugs, grasshoppers and caterpillars eating fruit and leaves on my plants. The birds also eat the bugs and caterpillars. Unfortunately my cat sometimes kills and eats grasshoppers and birds.

CONSUMERS
PRODUCERS

Exercise 5 Applying your knowledge

After reading Ecosystems on pages 159–161

Use what you know about living things and places to complete the table below.

Animal	Habitat	Food	Predators	Competitors	Climate and weather
seal					
	coral reef				
		microscopic plants and animals			
			spiders		
				other birds	
					very dry environment

Exercise 6 Analysing text

After reading Temperature and activity on page 161

Here are two paragraphs from the textbook.

Read them carefully several times and then answer the questions below.

> The body temperature of frogs changes with the outside temperature. Therefore, in the colder weather of winter, frogs hide away in logs or crevices.
>
> On the other hand, the body temperature of mammals (such as possums, wombats, mice, wallabies, bats and humans) and birds remains fairly constant all year round. These animals can be active all year round.

1 Find a *cause and effect linking word* in paragraph 1. (This word signals that the information after it is an effect of the information before it.)

____

2 Find a group of *contrast linking words* in paragraph 2. (These words signal that the information following will describe a difference from the information before them.)

____

3 Which words signal to the reader that a list will follow? _______________

____

4 Why do you think the editor of the textbook put brackets around the list of mammals? _______________

____

5 Which two main groups of animals have a constant body temperature? _______________

____

Fred Frog checks the outside temperature and, after careful consideration, decides that his best option is to go back to bed.

Exercise 7 Completing a table

After reading Survival—eat or be eaten on page 162

Apply what you know about the listed animals to complete the table below.

Animal	Features which help it get food	Features which prevent it being eaten
Haswell's frog		
spider		
giraffe		
octopus		

Exercise 8 Understanding words in context

After reading Ecosystems under threat on pages 165–167

You can often work out the meaning of an unknown word by reading the words around it (the context). Below are five words in italics which are used on these pages.

Use the context clues to try to work out their meanings.

The word in context	Your guess at the meaning (in your own words)
Over the past 40 years, there have been three major *outbreaks* of the crown of thorns starfish on the Great Barrier Reef.	
An arm that is *severed* from the starfish's body can regenerate into a new starfish.	
The maori wrasse, which are predators of starfish and were taken in large numbers by fisherman, are now *protected*.	
Ecosystems that have a wide variety of organisms tend to be able to *resist* changes such as diseases and severe weather conditions.	
It is estimated when 100 hectares of land are cleared, about 1500 birds die from *exposure*, starvation and stress.	

Exercise 9 Writing in role

Imagine you are a science reporter for a magazine. Write a short response (no more than 100 words) to each of the following questions from readers.

Dear Science Reporter,

I have read about the problems of the crown of thorns starfish on the Great Barrier Reef. Could you tell me what causes such a large population explosion of this particular animal?

Worried

Dear Worried,

____

____

____

____

Dear Science Reporter,

I have heard people talk about the word biodiversity but I don't understand what this means. Can you help me?

Puzzled

Dear Puzzled,

____

____

____

____

Dear Science Reporter,

I am 10 years old and I have to prepare a talk on the ways humans may have threatened the lives of many of the native animals and plants in Australia. Could you give me some ideas on what to write for my talk?

Talker

Dear Talker,

____

____

____

____

Roundup

Exercise 10 Devising possible solutions

After reading Biodiversity on page 166

Imagine you see a newspaper headline.

BIOLOGISTS INVESTIGATE KOALA DEATHS

Using what you have learnt about the survival of organisms in ecosystems, list possible reasons for the deaths of the koalas. The first one is done for you.

Factors	Possible reasons
Soil types and soil nutrients	The soil may have lost all the nutrients needed by gum trees to survive.
Weather conditions	
Competitors	
Predators	
Food availability	
Disease	

Exercise 11 Matching term and meaning

Match the term with its meaning by writing the letter with its number under the table.

Term	Meaning
1 Consumers	A Animals that eat plants or other animals
2 Producers	B A number of food chains together showing what all living things in a particular area eat
3 Biodiversity	C Microscopic organisms such as bacteria which cause dead organisms to decay
4 Photosynthesis	D Different groups of living things living in a community where eucalypt trees are most noticeable
5 First-order consumers	E A living place
6 Eucalypt community	F The process of getting energy from food
7 Food web	G Animals that eat producers
8 Respiration	H Different groups of plants and animals living together in the same place
9 Second-order consumers	I Any animal which hunts others as its prey
10 Scavengers	J All the different living things that form all the ecosystems on Earth
11 Predator	K The substance in plants which is able to absorb the energy in sunlight
12 Organism	L The complex system of feeding relationships and interactions between living and non-living things
13 Decomposers	M The food-making process in a green plant
14 Food chain	N Living things which make their own food from the sun's energy
15 Community	O Animals which eat the flesh and organs of dead animals
16 Ecosystem	P Any living thing
17 Chlorophyll	Q Animals that eat animals that eat producers
18 Habitat	R A sequence of living things in which each living thing is eaten by the next one

1 ______ 4 ______ 7 ______ 10 ______ 13 ______ 16 ______

2 ______ 5 ______ 8 ______ 11 ______ 14 ______ 17 ______

3 ______ 6 ______ 9 ______ 12 ______ 15 ______ 18 ______

Chapter 8
Magnetic forces

Overview

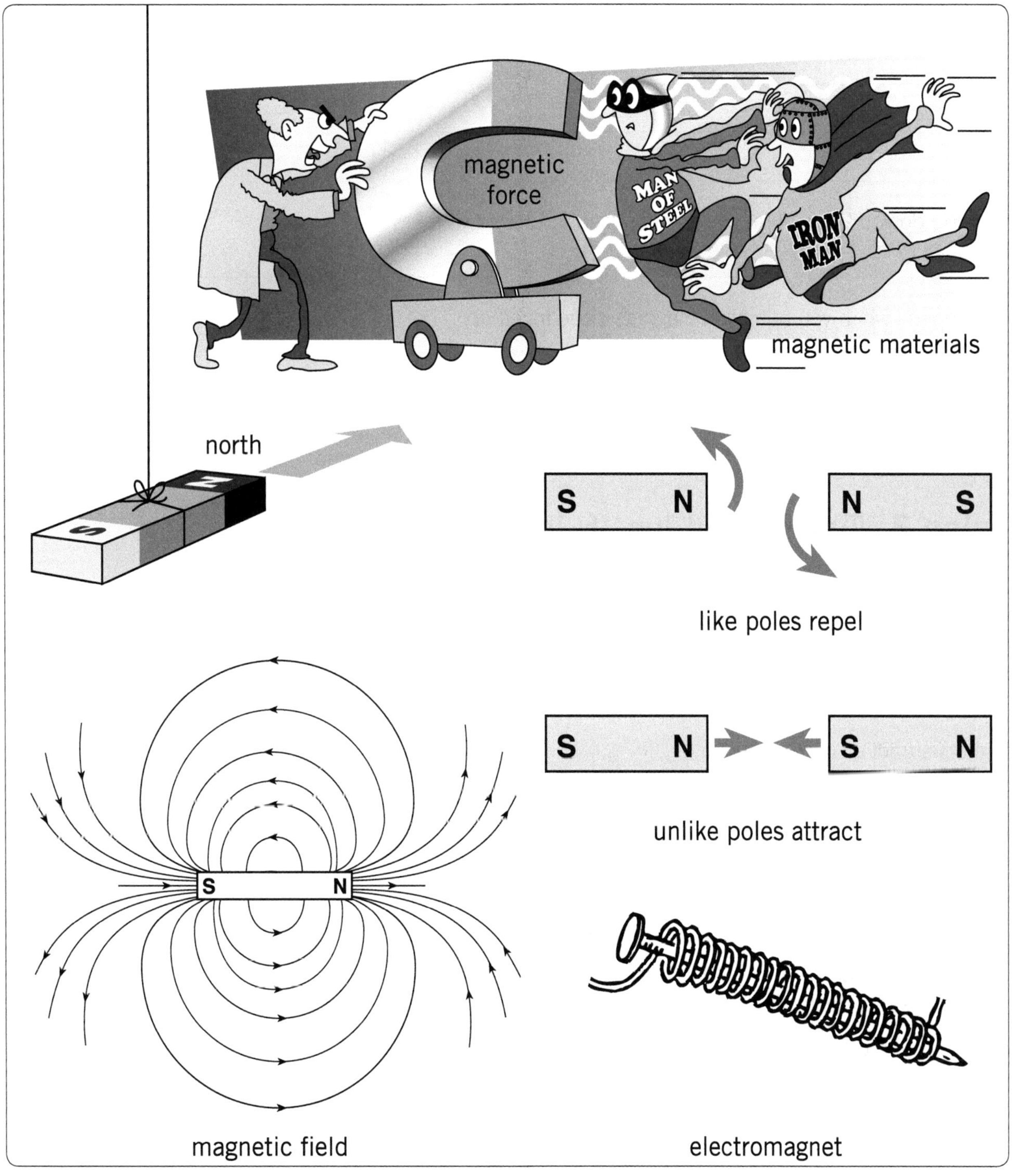

magnetic force

MAN OF STEEL

IRON MAN

magnetic materials

north

S N

N S

like poles repel

S N

S N

unlike poles attract

S N

magnetic field

electromagnet

What do you know already?

Exercise 1 Discussing in groups

In a small group, discuss each of the questions below. Write your answers in the spaces.

1 You have seen magnets holding important information on refrigerator doors. What force is holding these notices on? Why don't they stay on cupboard doors?

____

____

____

2 Where else do you know of magnets being used in everyday life?

____

3 What happens when you bring two magnets close together?

____

4 What is an electromagnet?

____

Exercise 2 Making and justifying decisions

Read the statements below carefully and then circle TRUE if you think the answer is true, and FALSE if you think it is false. Be prepared to give reasons for your answer.

1 A magnet can exert an invisible pull on a piece of iron without even touching it. TRUE FALSE

2 A magnet can be used to pick up coins. TRUE FALSE

3 Magnets attract all metals. TRUE FALSE

4 The ends of a magnet are more magnetic than the middle of a magnet. TRUE FALSE

5 Magnets will attract each other no matter which way they are facing. TRUE FALSE

6 A magnet suspended on a string will turn to point in an east–west direction. TRUE FALSE

7 A magnetised screwdriver will pick up the screws and nails. TRUE FALSE

8 You can make a temporary magnet by winding wire around a piece of iron and passing electricity through the wire. TRUE FALSE

9 A problem with electromagnets is that they cannot be turned on and off. TRUE FALSE

10 The Earth acts as if there is a giant magnet inside it. TRUE FALSE

Keep your answers in mind as you read the chapter.

During reading

Exercise 3 Completing a table

After doing Investigation 19 on pages 176–177

Complete the table by describing what you found out about magnets in each part of Investigation 19.

Investigation 19	Properties of magnets I found out
Part A	
Part B	
Part C	
Part D	

Exercise 4 Completing sentences

After doing Investigation 19 on pages 176–177 and reading Facts about magnets on page 178

Imagine you are a magician. How could you use the properties of magnets to appear to create 'magic'? Explain below using both words and diagrams.

Exercise 5 Interpreting an illustration

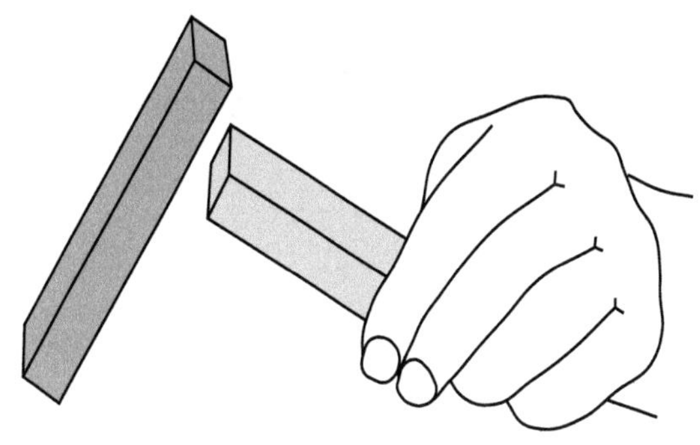

Samara has made a bar magnet sit as shown, without falling, using a second magnet. Explain how she has done this by marking the poles of the two magnets.

Is there more than one way to do it? Explain your answer.

Exercise 6 Completing cartoons and illustrations

After reading Magnetic fields on page 181

1 Complete the following cartoons in your own words.

2 Draw the 'invisible' magnetic field around the bar magnet on the right.

N S

3 Draw the combined magnetic field around the bar magnets on the right.

Exercise 7 Understanding on three levels

After reading The Earth's magnetic field on page 182

Understanding on three levels

For each of the statements below write T (true) or F (false) in the brackets provided.
When you have finished, discuss your answers in small groups. Try to reach agreement on the answers.
You should be ready to give reasons for your decisions.

Level 1: Reading for accuracy

(Be able to show where statements in the textbook support your answer.)

1 The south pole of Gilbert's imaginary magnet is near the Earth's South Pole. ( )

2 The magnetic North Pole is the same as the geographic North Pole. ( )

3 Magnetic compasses can be affected by variations in the solar wind. ( )

4 Auroras are beautifully coloured lights in the night sky. ( )

5 Sunspots are the origin of solar wind. ( )

Level 2: Drawing conclusions

(Be able to use information in the textbook to show how you arrived at this conclusion.)

1 There is a type of rock called magnetite which attracts things made of iron. ( )

2 The Earth's magnetism is caused by high-voltage underground power lines. ( )

3 The Aurora Australis can only be seen in northern Australia. ( )

4 In a place where compass needles swing wildly, radio signals would also swing wildly. ( )

5 Scientists have found evidence of electric currents in the molten core of the Earth. ( )

Level 3: Applying your knowledge

(Be able to apply your knowledge in a wider context.)

1 Charged particles are attracted by magnetic fields. ( )

2 If you heat a magnet in a burner flame it will lose its magnetism. ( )

3 At the North Pole, a compass needle free to pivot in a vertical plane will point straight down. ( )

4 Radio signals could not be transmitted around the world if there was no ionosphere. ( )

5 There could be no Aurora Australis if there was no ionosphere. ( )

Exercise 8 Analysing text and a diagram

After reading page 183

Below is an extract from your textbook. Read and examine it carefully, then answer the questions below.

Magnetic shielding

Magnetic fields pass through all substances except magnetic ones. Magnetic materials gather in, or concentrate, magnetic fields. These materials therefore act as shields, stopping the magnetic field from passing through them.

Look at Fig 3, which shows a piece of plastic and a piece of iron near a bar magnet. Notice how the magnetic field near the south pole is not affected by the piece of plastic—it passes right through it. At the north pole however, the field is gathered in by the piece of iron and does not go into the area above it. Anything above the piece of iron is *shielded* from the magnetic field.

Many sensitive scientific instruments need to be shielded from stray magnetic fields. This can be done by enclosing the object in a steel box.

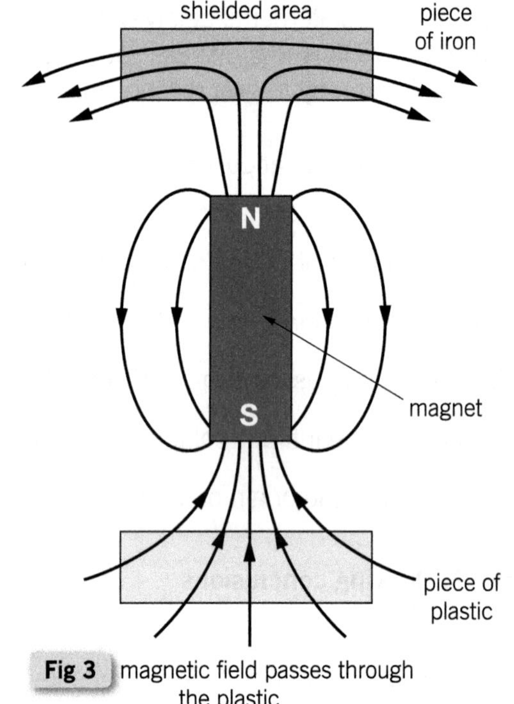

Fig 3 magnetic field passes through the plastic

1 In paragraph 1

a What is another word used, meaning 'gather in'? ______

b What materials are referred to by 'These materials'? ______

c Write the words which describe what 'shields' do ______

2 In paragraph 2

a Draw an arrow from the diagram to the sentence which describes what happens to a magnetic field in a piece of plastic.

b Draw an arrow from the diagram to the sentence which describes what happens to a magnetic field in a piece of iron.

3 In paragraph 3

'This' refers back to:

a sensitive scientific instruments

b stray magnetic fields

c shielded.

Rewrite the last sentence in paragraph 3 by substituting the correct words for 'this'.

______

______

______

Exercise 9 Predicting and inferring

After doing Investigation 21 on page 184

Ahmed was experimenting with an electromagnet similar to the one you made in Investigation 21. His electromagnet was made of 40 turns of wire around a large nail. He connected the wires to a power supply set on 2 volts. He found that, on average, this electromagnet picked up 10 paperclips.

Ahmed tried different voltages through the electromagnet and counted the number of paperclips it could pick up each time. His results are shown in the data table on the right.

Voltage	Average number of paperclips picked up
1	5
2	10
4	20
5	25
8	40

1 Predict how many paperclips could be picked up if the power supply was set on 3 volts. Explain how you arrived at your answer.

____

____

____

2 Use the data to draw a line graph.

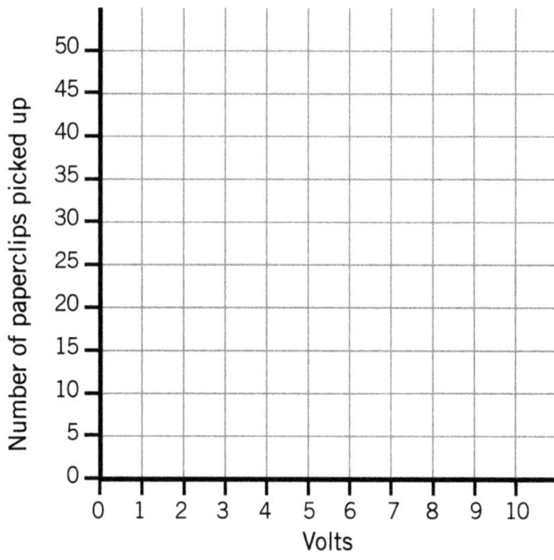

3 Predict how many paperclips could be picked up if the power supply was set on 10 volts.

____

4 Kristy made an electromagnet with the same sort of wire as Ahmed used. She set the voltage to 4 volts and found that she could pick up, on average, 36 paperclips. Infer how Kristy's electromagnet was different from Ahmed's.

____

____

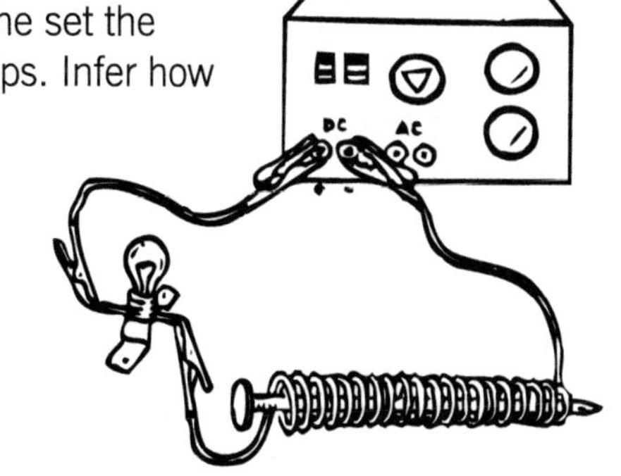

Exercise 10 Rearranging the sequence

After reading Uses of electromagnets on pages 189–190

There are seven steps that occur when you speak into the mouthpiece of a telephone and someone else listens on the earpiece of another telephone. These steps are listed on the next page, but they are in the wrong order. Rewrite them in the correct order.

Electric currents pass through the electromagnet in the earpiece.

Electric currents travel along telephone wires.

The metal disk vibrates as the electromagnet switches on and off.

Sound waves are changed to electric currents in the mouthpiece.

Sound waves travel to your ear.

You create sound waves when you speak.

Vibrations produce sound waves.

1 You create sound waves when you speak.

2 ______________________________

3 ______________________________

4 ______________________________

5 ______________________________

6 ______________________________

7 ______________________________

Exercise 11 Relating cause and effect

After reading Uses of electromagnets on pages 189–190

In everday life and in science you often give the reason for one thing leading to another. Suppose you have a toothache. This is the effect (what occurs). The cause of this is a decayed area in one of your teeth. In this way you relate the cause (the reason) to the effect.

Use the information in the boxes on pages 189–190 to fill in the missing **causes** and **effects** in the table below.

Cause	Effect
The current through the electromagnet crane is turned off.	
	The metal disk in the telephone earpiece vibrates.
A person wearing a stud belt walks through the airport metal detector.	
	The maglev train starts, stops or changes its speed.

Exercise 12 Using cause and effect linking words

In this chapter, you learnt that some actions cause another action to occur.
For example, if the current through the electromagnet crane is turned off, the load would drop.

You can write about this cause and effect using *cause and effect linking words*.

Cause and effect linking words (for actions that cause another action to occur)

therefore	so that	an outcome of	results in
due to	as a result of	causing	consequently
since	because	caused by	this leads to
the effect of	the reason for	if … then	when … then
resulting in	making	as	accounts for
so	thereby	causes	gives rise to
creates	owing to	hence	thus

For example

- **Because** the current through the electromagnet is turned off, the load drops.
  cause: the current through the electromagnet is turned off; effect: the load drops.
- The load drops **as a result of** the current through the electromagnet being turned off.
  effect: The load drops; cause: the current through the electromagnet being turned off.

Use *cause and effect linking words* to describe these actions and results.

1 North poles of two magnets together ⟶ repel each other

____

____

2 Electric current flowing through wire ⟶ magnetic field

____

____

3 Magnetic field passing into a piece of iron ⟶ magnetic shield

____

____

Roundup

Exercise 13 Writing in role

Work in pairs. You will need a tape recorder to record an interview, or sit behind a screen to present your interview to the class.

One of you is a famous professor and your special area of expertise is magnetism.

Think of a good name for yourself. Professor ______________________

The other person is a radio compere who knows very little about magnetism.

Think of a name for yourself. ______________________

Record an interview on magnets and magnetism which starts and finishes like this:

Compere: Good evening listeners. Welcome to station __________. This is your favourite radio host, ______________________, and I am very pleased to have with me here tonight the world famous expert on magnetism, Professor ______________________. Welcome Professor.

Professor: Thank you, ______________________.

Compere: Now, Professor, tell us a few things about magnetism. First of all, what is magnetism?

Professor: ____

____

Compere: ____

____

Professor: ____

____

Compere: ____

Professor: ____

____

Compere: ____

Professor: ____

____

Compere: Thank you, Professor, for sharing some of your expert knowledge with us. I'm sure the listeners out there are much better informed than they were before. Be sure to be listening next week when I will have the pleasure of interviewing ______________________ who is an expert on ____.

Chapter 9
Consumer science

Overview

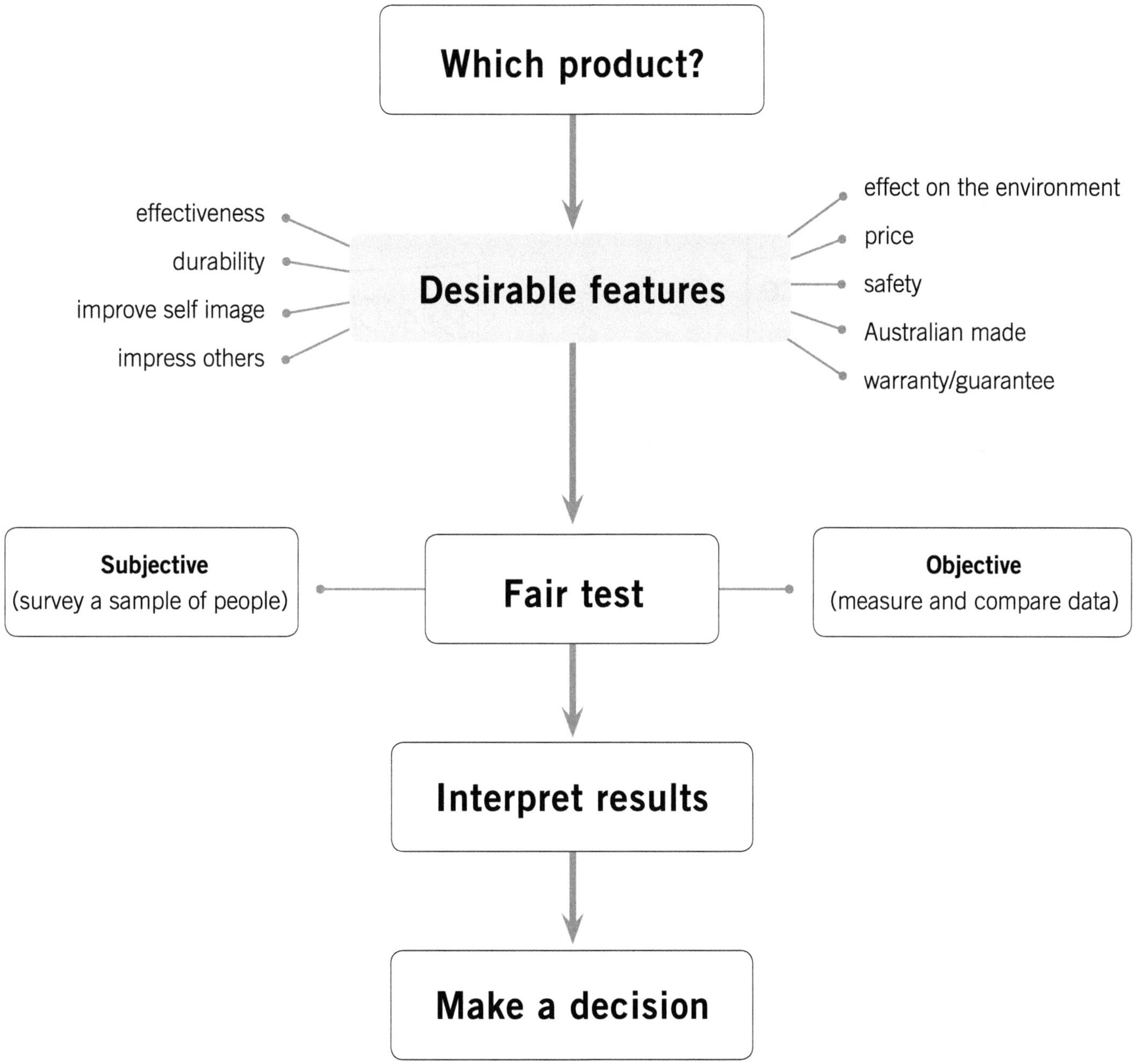

Which product?

↓

Desirable features

- effectiveness
- durability
- improve self image
- impress others
- effect on the environment
- price
- safety
- Australian made
- warranty/guarantee

↓

Fair test

- **Subjective** (survey a sample of people)
- **Objective** (measure and compare data)

↓

Interpret results

↓

Make a decision

What do you know already?

Exercise 1 Comparing and contrasting

1 Your grandfather has been told by his doctor that he must reduce his intake of salt, sugar and food preservatives. He sends you to the shop to buy some tomato sauce for him. You examine the labels on four brands of tomato sauce very carefully before making your purchase.

FREE FROM PRESERVATIVES

HEALTHY

TOMATO SAUCE

....best for you and your health.
Made from the finest tomatoes.
Satisfaction guaranteed or money back.

500 mL

Ingredients: Organic Tomatoes, Salt, Vinegar, Spices

LOW IN SALT

Homestyle

Tomato Sauce

Ingredients per 100 g: Tomatoes 95 g, Salt 1.5 g, Sugar 3 g, Food Acid (260) 0.5 g

2 Which one would you buy? Explain in detail why you made this choice.

3 What other food information do you think is missing from the labels, that would allow you to make a more informed decision?

4 Which of these two washing powders is better value for money? (Assume both powders are equally effective.)

Your selection—Ultra Bright or Super Wash? ______________________

Show your working.

5 An advertisement for Atomic Super Bath and Tile Cleaner says that it cleans brighter, faster and saves you money because it uses less.

a The makers of Atomic Super Bath & Tile Cleaner state three claims. What are they?

____

____

____

b How could you show that Atomic Super Bath & Tile Cleaner uses less?

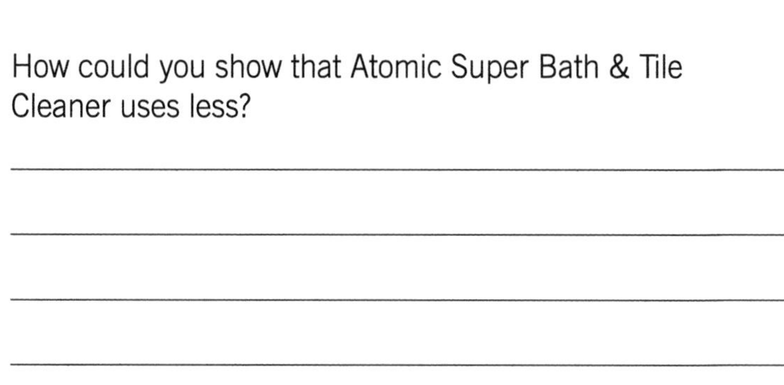

____

____

____

____

During reading

Exercise 2 Interpreting a table

After reading Consumer testing on pages 197–199

1 What is a disadvantage of the highly recommended Brand A phone?

____

____

2 What are two good features of the lowly ranked Brand H phone?

____

3 What does 'ergonomics' mean?

____

4 List a phone which disproves the generalisation: the more expensive the item, the higher the quality.

____

5 Which phone would you purchase? ______________________________

Give reasons for your choice. ______________________________

____

____

Exercise 3 Designing a survey

After reading pages 199–200

You are a manufacturer of backpacks. You decide to expand into the teenage market. Your company has always made backpacks in black only, but your marketing consultant thinks that you would sell more backpacks to teenagers if they were produced in brighter colours. You decide to gather more information by doing a survey.

1 Which group(s) would you survey? ______________________________

2 What size sample would you use? ______________________________

3 In which location(s) would you survey? ______________________________

4 In groups, design a simple survey questionnaire of at least five questions so that the person asking the questions only needs to put a circle around words already on the questionnaire, as shown below.

a Which age group are you? Under 12 years (12–19 years) 20–30 years

b ______________________________

c ______________________________

d ______________________________

e ______________________________

Exercise 4 Interpreting a label

After reading Option 4—Food additives on pages 210–211

Examine the label below very carefully. Answer the questions and rule a line from each question to the place on the label where you found the answer.

1 What is the datemark?

____

2 What is the name of the manufacturer?

____

____

3 What is the mass of the product?

____

4 How many kilojoules of energy are there in a 200 g serving?

____

5 Does it contain an artificial sweetener?

____

6 Where would you write to complain about the product?

____

7 What colour is the product?

____

8 What is the product's name?

____

____

9 Which additive was used to thicken and set the product?

____

____

Exercise 5 Reading the fine print

In order to promote their products to a health-conscious society, advertisers stress the healthy elements of their product (even though we are the heaviest generation ever!).

While Australian advertising standards ensure that what is stated is generally accurate, consumers need to read the fine print on labels.

Examples:

- a product is labelled as '92% fat-free' but the label shows 8 g of fat per serving.
- a product is advertised as 'low in fat' but the label shows it is very high in sugar content.
- a product is advertised as 'extra light' but the extra light refers to colour, not ingredients.

Look through supermarket shelves and see if you can find more examples of misleading information.

Exercise 6 Examining an issue from different perspectives

Debate is currently taking place in Australia about buying Australian made and owned products.

For If Australians buy products made in Australia and the profits stay in Australia, then more wealth and jobs are created and the future is better for future generations.

Against The Australian economy depends on other countries buying Australian products—natural resources, manufactured goods or services. Foreigners will not be keen to continue buying Australian if they see Australians rejecting foreign products, so they will find other sources for the goods they used to buy from Australia. Without foreign investment and foreign trade, Australia would be poorer and have high unemployment.

What do you think?

Notes: ______________________________

Roundup

Exercise 7 Understanding persuasive techniques

Advertising agencies use many techniques to persuade the consumer to purchase a particular product. Some of these techniques are listed in the table below. In the second column, list some products that you associate with this advertising technique.

Advertising Technique	Product
Association with luxury	
Price/value for money	
Safety	
Australian product	
Peer pressure	
Youth, fun, friends	
Earn the chance to win money or expensive items when you purchase this product	
Health	
Scientific facts	

Exercise 8 Applying persuasive techniques

In pairs, choose one of the advertising techniques above and write your own radio advertisement for an imaginary product. Also target a particular sex, age group or economic group. Make sure you are very persuasive so that the consumers in your target group don't stand a chance.

Chapter 10
How things work

Overview

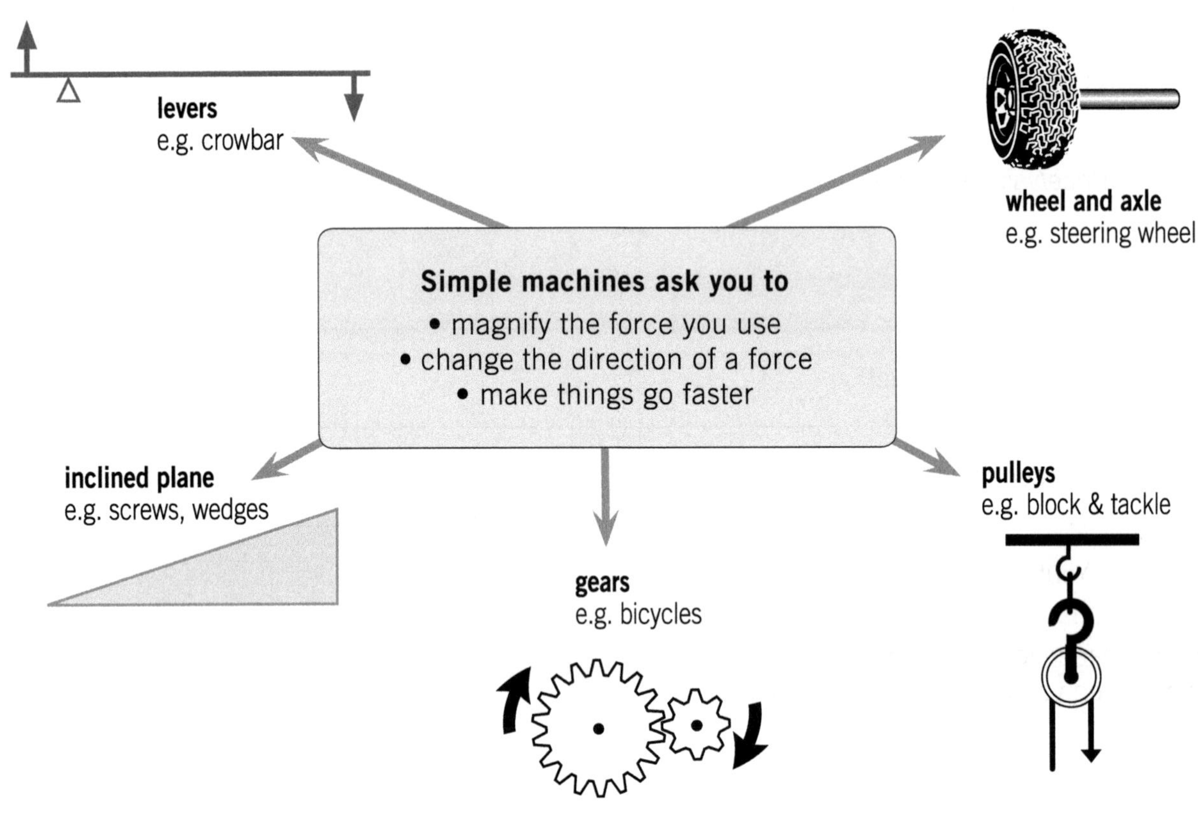

Flying machines work by action and reaction

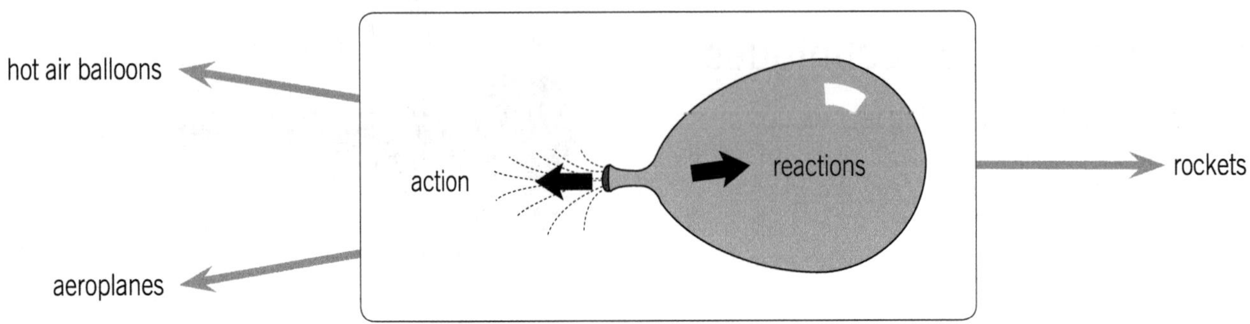

Exercise 1 Matching roles and responsibilities

In group work, each member's contribution is of equal worth. Sometimes you will need to take on a role within your group. The six roles and six role descriptions below have been jumbled. Match each role with its responsibilities. Then, as a whole class and with your teacher's guidance, come to an agreement so that you all have the same understanding of the responsibilities of the different roles.

Role	Role description
Encourager	has the job of informing the rest of the class of the findings, conclusions and results of the group's efforts. Gives credit where it is due, and names group members whose ideas the group has adopted, e.g. (Name) suggested … and (another name) added …
Reader/Checker	has the job of moving the group through the topic. Uses phrases such as: "Has anyone got any ideas to begin with?" "Can anyone add to that?" "That was a good idea (person's name)." "What do you think about that idea (person's name)?"
Recorder	has the job of judging the group's performance and the roles of each group member. This can be done by comparing the particular group's behaviour with the characteristics of effective and ineffective groups on page 87 of this workbook. Uses phrases such as "Our group will be better when …" "We need to …" "A good feature of the discussion was that …"
Reporter	has the job of ensuring that the topic to be explored is done in the time given. Uses phrases such as: "We should be finished by …" "It's half-way through our time. How should we spend the rest of the time?" "We only have a minute left."
Timer	has the job of making sure that everyone is clear about the task to be done. This means reading the task to the group, ensuring the group stays on task, and checking that everyone agrees before the recorder records the group's decision. Uses phrases such as: "We have to …" "That's not what we're being asked to do." "Are you all happy about that?"
Evaluator	has the job of recording the group's decisions in a way that the group agrees upon and that can be read by someone else.

Exercise 2
Devising possible solutions

Problem 1: Zachary is using a piece of wood to move a large rock in the garden, but it won't budge. Suggest two things Zachary could do to make it easier to move the rock.

1 ______________________________

____

2 ______________________________

____

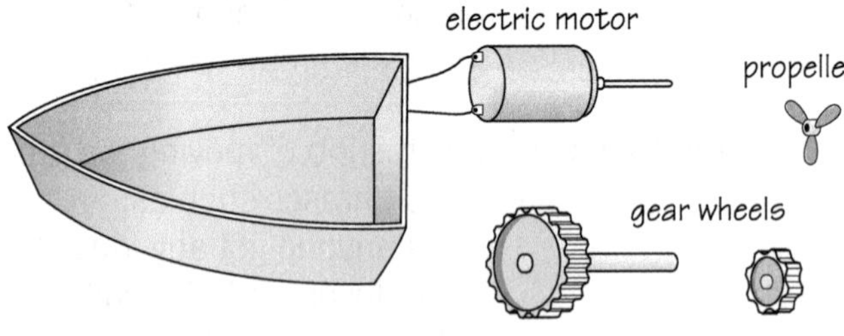

Problem 2: Selina has made a model boat using a small electric motor and a propeller. When she tests it out in the bath she finds it goes too fast. Her father gives her two gear wheels—one bigger than the other—and suggests she uses these. Draw a diagram showing how Selina could put the various parts together to make her boat go more slowly.

Keep your solutions to these two problems in mind as you read and work your way through the chapter. This way you will find out which principles of physics you already know.

Your solution:

During reading

Exercise 3 Completing sentences

After reading Simple machines on page 221

Fill in the missing words in the sentences below.

1 Machines help you to do things ______________________________

2 The most common simple machines are levers, ______________________________

____

3 The larger the hammer ______________________________ the force you need to apply.

4 A pulley helps you by changing the ______________________________ of the force.

For example, to raise a flag up the flagpole you ______________________________

5 Machines that make things go ______________________________ are said to have a

____

Exercise 4 Recalling information

After reading Levers on page 222

1 Lorenzo is part of an acrobatic troupe. To propel his brother high into the air, he has to jump on the end of a plank as shown in the cartoon.

a Put the following labels on the cartoon. Use arrows to show direction.

lever   pivot   load   effort

b Which is larger—the load or the effort?

Why? ______________________________

2 You want to shift a large rock using a wheelbarrow. Would it be better to put the rock in the front of the wheelbarrow or at the back? ______________________________

Explain your answer. ______________________________

Exercise 5 Interpreting a diagram and data

After doing Investigation 27 on page 223

Mario did PART A of Investigation 27 as shown.

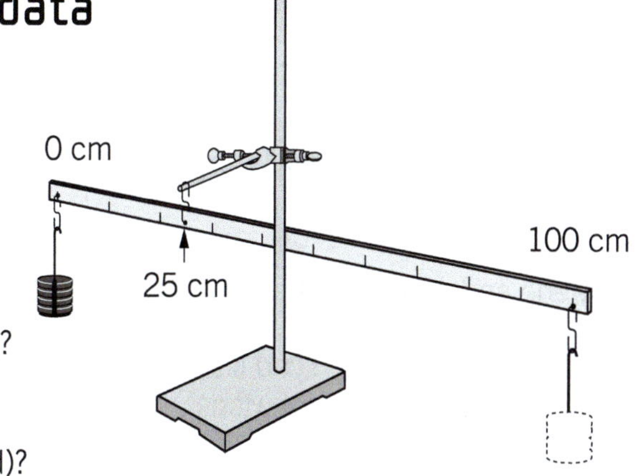

1 On the diagram mark the load, the effort and the pivot.

2 What is the load? ________ grams

3 What is the load distance (the distance the load is from the pivot)?

____

4 What is the effort distance (the distance from the pivot to the load)?

____

Mario recorded his results in a data table.

Load (N)	Load distance (cm)	Effort (N)	Effort distance (cm)	Load effort
2	5	0.11	95	18
2	15	0.35	85	6
2	25	0.67	75	3
2	35	1.08	65	2
2	50	2.00	50	1
2	70	4.67	30	0.4

5 What effort was needed for the set-up in the diagram?

____

6 Complete the missing words in Mario's generalisations.

As you move the pivot further from the load, the effort needed ______________________________

As you move the pivot closer to the load,

____

7 The load divided by the effort is called the load–effort ratio. What does a load–effort ratio of 1 tell you? ______________________________

8 If the load in the diagram was 600 g, what effort would be needed? ______________________________

Explain your prediction ______________________________

Exercise 6 Correcting information

After reading Other simple machines on pages 224–225

Each of the following statements is incorrect. Rewrite each one correctly.
You only need to change one or two words.

1 A ramp is a simple machine called a wheel-and-axle.

____

2 The steeper the ramp the easier it is to roll the barrels onto the truck.

____

3 To use a jack you use a large force to lift a small load.

____

4 The smaller the steering wheel the easier it is to turn it.

____

5 A screwdriver is an example of a wheel-and-axle. The handle is the axle and the blade is the wheel.

____

Exercise 7 Analysing data

After reading Pulleys and Gears on page 227 and doing Investigation 28

Raylene and Jason did the investigation and recorded their results in a data table as shown.

Number of pulleys	Load lifted (newtons)	Distance load moves (cm)	Effort needed to lift (newtons)	Distance effort moves (cm)
1	2	50	2.1	50
2	2	50	1.2	100
4	2	50	0.7	200

1 What was the reading on the spring balance when they used one pulley to lift a load of 2 newtons (200 grams)?

____

2 When they used two pulleys, how far did they have to pull the rope? ______________________________

3 Look at the data for the four-pulley system. Fill in the missing words describing what happened.

With four pulleys the effort needed to lift a load of ______________ a distance of ______________

was ______________ They had to move the spring balance ______________________________

4 Raylene predicted that the effort needed using one pulley would be exactly 2 newtons. Suggest why the force they measured was slightly more than this (2.1 N). Write a complete sentence.

____

Exercise 8 Interpreting illustrations

After reading Gears on page 229

Examine the gears in 1 and 2 on the right of page 229

1 Count the number of teeth on the gear wheels and record this in the table below. Then go to question 2.

	Motor speed (revolutions per minute)	Number of teeth on driving gear	Number of teeth on driven gear	Gear ratio	Propeller speed (revolutions per minute)
1	200				
2	200				

2 The gear ratio can be found as follows:

$$\text{gear ratio} = \frac{\text{number of teeth on driving gear}}{\text{number of teeth on driven gear}}$$

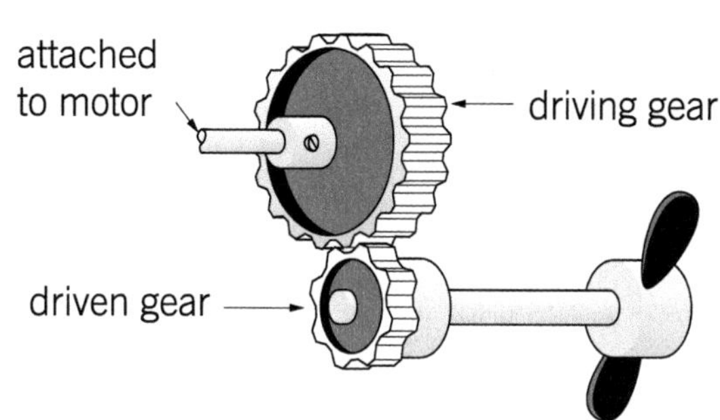

So in 1 the gear ratio = =

You can find the propeller speed as follows:

propeller speed = motor speed x gear ratio

= 200 ____________

= ____________ revolutions per minute

Add your answers to the table above.

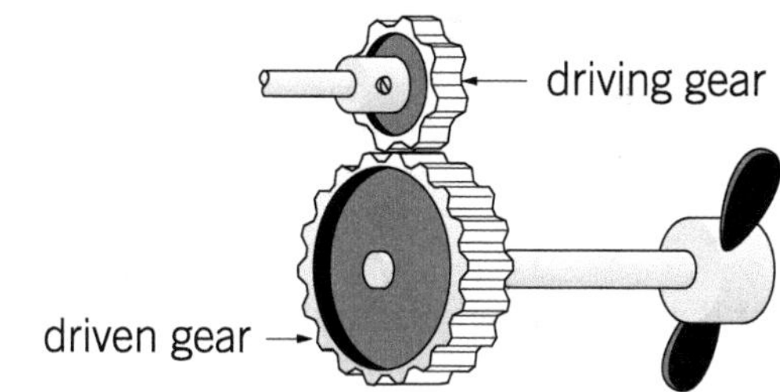

3 Calculate the gear ratio and propeller speed for 2. Add your answers to the table.

4 Complete this generalisation.

If the gear ratio is greater than 1, the propeller turns ____________ than the motor. If the gear ratio is less than 1, ____________.

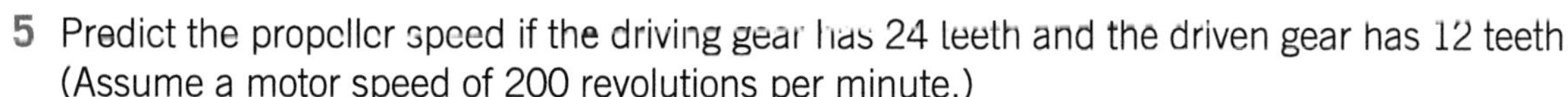

5 Predict the propeller speed if the driving gear has 24 teeth and the driven gear has 12 teeth. (Assume a motor speed of 200 revolutions per minute.)

Space for working

Exercise 9 Interpreting a table and an illustration

After reading The bicycle on page 230 and Gear ratios on page 232

A mountain bike has the following gearing system. The table on the left shows the number of teeth on each sprocket on the rear wheel.

Gear	Number of teeth on sprocket
1	28
2	24
3	20
4	18
5	16
6	14

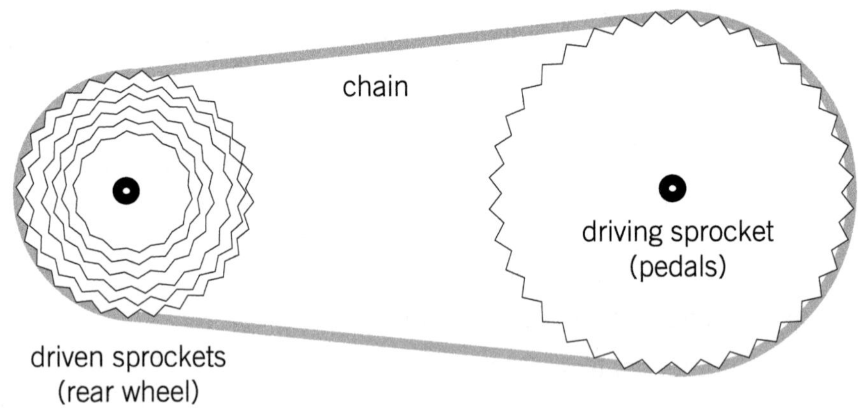

1 Which gear would you use for travelling up a steep hill? ______________

2 Which gear would you use for travelling fast on a flat road? ______________

3 Gear ratio $= \dfrac{\text{number of teeth on driving sprocket}}{\text{number of teeth on driven sprocket}}$

a Using this formula, find the gear ratio for first gear.

= ______________

b What is the gear ratio for sixth gear?

= ______________

4 How much faster is sixth gear than first gear (for the same pedal speed)?

____

____

Exercise 10 Completing a table

After reading Hot air balloons on pages 235–236

Look at the diagrams of hot air balloons below. In each case the arrows represent the forces acting on the balloon. The size of the force is indicated by the length of the arrow.

In the middle column indicate whether the forces on the balloon are balanced or not. In the right-hand column briefly describe what happens in each situation.

Situation	Are the forces balanced (✔ or ✘)	What happens?
force from rising hot air gravitational force		
force from wind		

Exercise 11 Labelling and predicting

After reading Rockets and Aeroplanes on pages 236–238

1 Ian is learning how to fly. On the arrows in the drawing, write in the four forces listed below.

frictional drag
gravity
lift
thrust

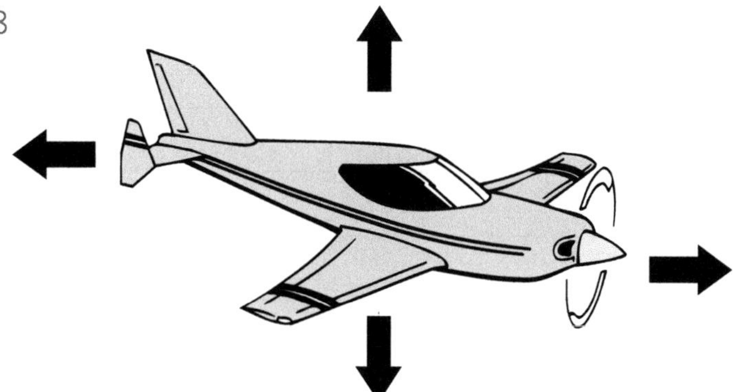

2 Are the forces on the plane balanced or unbalanced? ______

3 The plane is presently travelling at a constant speed at a constant height.
What would happen if the lift decreased?

______

4 Suggest two different ways of slowing down the plane.

a ______

b ______

Exercise 12 Sequencing

After reading Rockets and Aeroplanes on pages 236–238

Place the following sequence in the correct order by placing the numbers 1–7 in the boxes

1 ☐ Air rushes over the wings.
☐ The plane lifts off the ground.
☐ The faster-moving air creates a lower pressure than slower-moving air.
☐ The upward force is greater than gravity.
☐ The difference in air pressure creates an upward force.
☐ The plane speeds up along the runway.
☐ The air moves faster over the top of the wing than the bottom of the wing.

2 Use the seven sentences in the list above and some *cause and effect linking words* from page 92 of this workbook to write a paragraph explaining how an aeroplane takes off.

______

⮂ Roundup

Exercise 13 Crossword

Check your knowledge of the key words in the chapter by doing this crossword. (Where the answer is more than one word, don't put a space between the words.)

Across clues

2 Slanting surface used as simple machine

4 Grooved wheel with a rope around it

7 Point on which a lever turns

8 Wheels with teeth on them

12 For every action there is an equal and opposite ________

14 Simple machine consisting of wheel attached to axle

15 Force which pushes rocket upwards

Down clues

1 Helps you do things more easily

3 Force needed to do something

5 Rigid bar which rotates about a point

6 Downwards force

9 Gear wheels on a bicycle

10 Frictional force which acts on an aeroplane

11 Special type of inclined plane that spirals round and round

13 Useful simple machine when you have a flat tyre

Chapter 11
Earth's resources

Overview

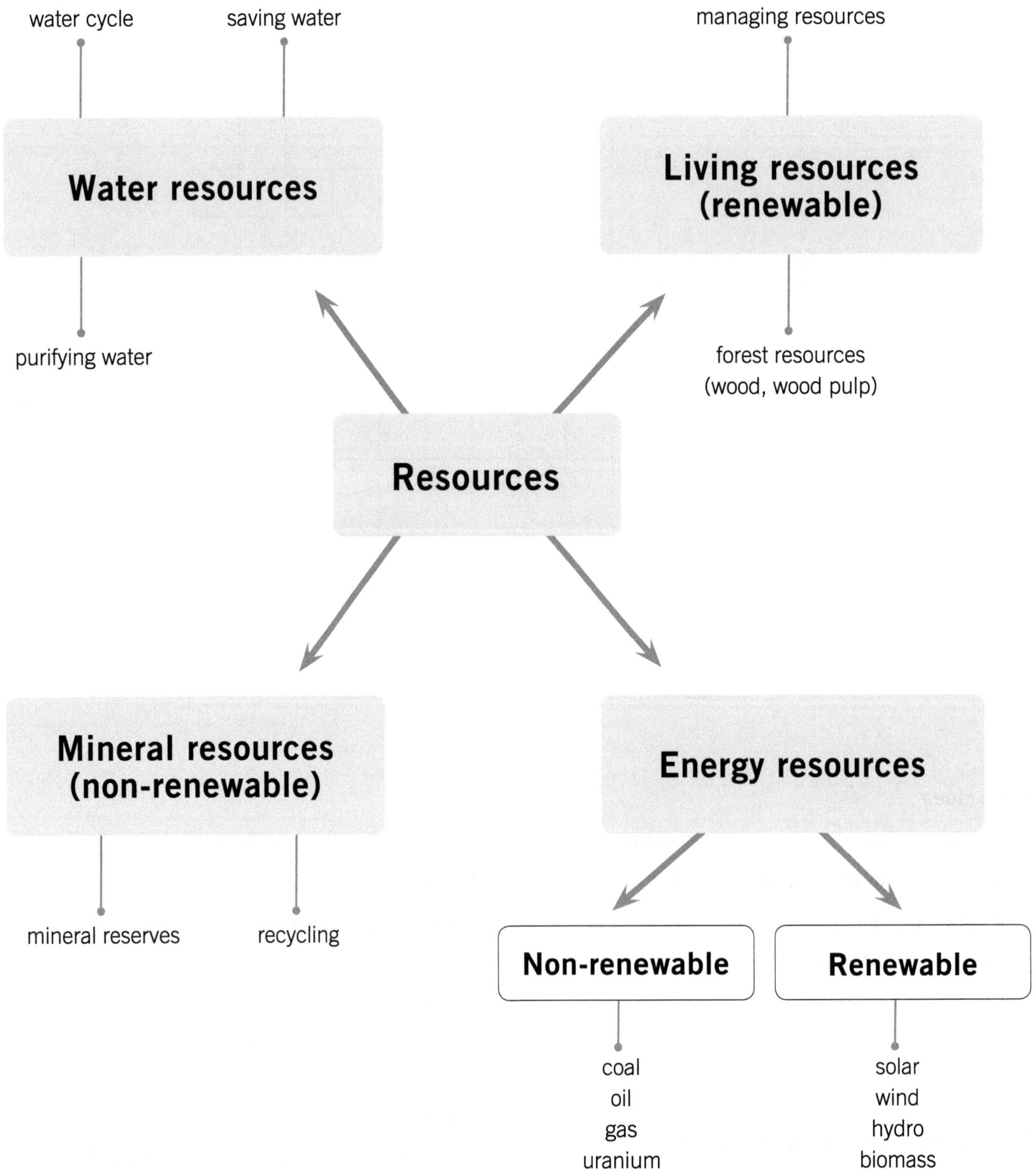

Resources

Water resources
- water cycle
- saving water
- purifying water

Living resources (renewable)
- managing resources
- forest resources (wood, wood pulp)

Mineral resources (non-renewable)
- mineral reserves
- recycling

Energy resources
- **Non-renewable**: coal, oil, gas, uranium
- **Renewable**: solar, wind, hydro, biomass

Exercise 1 Working in groups

In this chapter, you will be doing the activities and exercises in small groups. Working in groups is a very practical and co-operative way of generating ideas and using language to learn. Before commencing the exercises, you need to understand what makes an effective group, and the roles and responsibilities of group members.

1 Read through the lists of characteristics of **effective** and **ineffective groups** below. They have been organised according to the behaviour of their members.

2 Circle the *three* characteristics of **effective groups** that you think are the most important for successful group work in your classroom.

3 Now circle the *three* characteristics of **ineffective groups** that you personally need to keep in mind when you work in a group.

Characteristics of effective and ineffective groups

Effective groups

1 The atmosphere tends to be informal and comfortable, with group members sitting so that all group members are facing each other.

2 There is a lot of discussion in which everyone takes part. Everyone keeps to the point.

3 Everyone understands the task they have to do.

4 The group members listen to each other. Every idea is given a hearing.

5 When there is disagreement, the group is comfortable with this and works towards sorting it out. Nobody feels unhappy with decisions made.

6 People feel free to criticise and say honestly what they think. Group members show that a contribution is valued by using encouraging body language such as nodding, smiling and eye contact; and by giving credit and using first names of people who contribute good ideas.

7 Everyone knows how the other members of the group feel about what is being discussed.

8 When action needs to be taken, everyone is clear about what has to be done, and they help each other.

9 Different people take over the role of leader from time to time.

10 The group knows how well it is working and what is interfering with its progress. It can look after itself.

Ineffective groups

1 The atmosphere reflects indifference or boredom, with little eye contact between group members.

2 Only one or two people talk. Little effort is made to keep to the point of the discussion.

3 Group members are unsure of the task.

4 People do not listen to each other. Some people do not put their ideas forward to the group.

5 Disagreements are not dealt with effectively. They are put to the vote without being discussed thoroughly. Some people are unhappy with the decisions.

6 People are not open about what they are thinking as they fear their ideas will not be valued by others.

7 One or two people are dominant. They ignore other members' feelings.

8 Nobody takes any interest in what has to be done, and nobody offers to help others.

9 Only one or two people make the decisions and act as group leaders.

10 The group does not talk about how it is working or about the problems it is facing. It needs someone to look after it.

What do you know already?

Exercise 2 Discussing

Form groups of four to six to discuss and report on the following topics. Report your discussion decisions to the rest of the class.

1 List the ways in which you use water each day in the home.
Make the list in order of *uses most* to *uses least*.

2 List your group's four suggestions for saving water in the home.

3 List five things that can be used or made from trees from a forest.

4 List five minerals that are mined in Australia.

5 What difference does recycling make to our use of resources?

6 List the problems that will arise from our use of fossil fuels.

Working space

During reading

Exercise 3 Describing

After reading Purifying water on page 248

Complete the table below by describing the action or use of each component in a water treatment plant.

Water treatment plant component	Action or use
Reservoir	
Alum and mixing tanks	
Settling tanks	
Filtration	
Chlorine	
Water storage	

Exercise 4 Sequencing

After reading Purifying water on page 248

When you are asked to describe actions which occur one after another, you should use some of the *sequence linking words* on page 92 of this workbook.

Use the information on page 248 and in the table you completed in Exercise 3 to complete this paragraph describing how water is purified for drinking.

Water that is stored in a reservoir is piped into the water treatment plant. To begin with, alum is added

Exercise 5 Completing a table

After reading Resources from plants on page 253

Complete the table below by describing the product that each plant resource is used for or made into.

Product from plant resources	Product is used to make ...
Fibres	
Building materials	
Food	
Recreation, decoration, beauty	
Fuel	
Medicines	

Exercise 6 Interpreting information

After reading Australia's mineral reserves on page 257

Look at the table on page 257 and answer the following questions.

1 Which metal resource had the largest reserves in 2009? ______________________

2 Which metal resource had the smallest reserves in 2009? ______________________

3 Which metal resource is being mined at the greatest rate? Justify your answer. ______________________

____

____

4 Look at the data for black coal and brown coal. Which one will last the longest? ______________________
Justify your answer, and explain the assumptions you had to make about the reserves of both types of coal.

____

____

____

____

Exercise 7 Explaining information in a table

After reading Energy resources on page 259

Look at the table on page 259 and answer the following questions.

1 Explain how the information in the table is organised ______________________________

____

____

2 Suggest why the percentage use of uranium is 0%? ______________________________

____

3 Which energy resource is mainly used for transportation? ______________

4 Which renewable energy resources could be used for transportation? ______________

5 Explain how your suggestions in **4** could be used. ______________________________

____

____

Roundup

Exercise 8 Conducting a class quiz

1 Select a quiz master and a scorer, then divide the rest of the class into three equal teams (Team A, Team B and Team C).

2 Distribute a blank card to each student. On the card, each student is to write a Chapter 11 question (and an answer on the other side).

3 Team A students place their cards in the Team A box.
Team B students place their cards in the Team B box.
Team C students place their cards in the Team C box.

4 The quiz master selects a question card from the Team A box and asks the question of Team B.

5 If Team B members get the answer correct, the scorer awards one point to Team B. If the answer is incorrect, the team doesn't score.

6 The quiz master then selects a question card from the Team B box and asks the question of Team C.

7 If Team C members get the answer correct, the scorer awards one point to Team C. If the answer is incorrect, the team doesn't score.

8 Next, the quiz master then selects a question card from the Team C box and asks the question of Team A. The score is recorded.

9 When all question cards have been selected, the team with the highest score is the winner.

Linking words

Sequence linking words (for actions that occur one after the other)

first, second	then	before	finally	subsequently
to begin with	later	next	gradually	initially
when	from there	last	initial	after

Cause and effect linking words (for actions that cause another action to occur)

therefore	so that	an outcome of	results in
due to	as a result of	causing	consequently
since	because	caused by	this leads to
the effect of	the reason for	if … then	when … then
resulting in	making	as	accounts for
so	thereby	causes	gives rise to
creates	owing to	hence	thus

Contrast linking words (to describe differences)

on the other hand	in contrast to	different from	however
alternatively	but	although	rather than
whereas	while	unlike	yet
even though	less than	more than	whilst

Comparison linking words (to describe similarities)

both	same	in both	like
in both cases	similarly	and	as well as
in the same way	just as … so …	alike	similar in that

Generalisation linking words (for statements that are true in most cases)

generally	usually	commonly	many
normally	as a general rule	most	mostly